SUSTAINABILITY PERCEPTION INDICATORS

This book offers a comprehensive analysis of the background, practice, potential and challenges associated with developing and using perceptual indicators for assessing sustainability.

Sustainability Indicators (SIs) are usually described and portrayed as quantitative metrics that rest above the human realm of subjectivity, opinion and bias. Thus, they are the basis for objectively measuring progress towards the attainment of targets in sustainable development. Unlike the "hard" metrics described above, the way in which we experience the world and frame our personal decisions based on that experience, past and present, is founded on perceptions, and these can be inherently subjective. This book argues that perception-based indicators are an important subset of SIs in assessing sustainability, environmental quality and well-being. The chapters draw upon examples such as the Corruption Perception Index (CPI), produced by Transparency International (TI), and a variety of happiness indices, amongst many others. The chapters in the book can be thought of as comprising two parts. The first provides a broad review of the field of SIs, addressing the challenges involved in identifying, defining and populating SIs, especially for countries in the Global South. The second part of the book summarises the range of participatory approaches that have often been used to develop perception-based indicators of sustainability, along with their respective pros and cons.

This book will be useful for students studying social sciences, economics, environmental studies, human geography, politics and international development. The book may also appeal to students taking courses in business studies.

Stephen Morse holds the position of Chair in Systems Analysis for Sustainability at the Centre for Environment and Sustainability, University of Surrey. He has a background in applied ecology and the environment, and his research and teaching interests are broad, spanning both the natural and social sciences. These interests include methods for the assessment of sustainability (e.g. indicators and indices) to help guide intervention. He has also helped pioneer some participatory methodologies for sustainability assessment. He has been involved in research and sustainable development projects across Europe, the Mediterranean, Africa and Asia.

Routledge Studies in Sustainability

Digital Technologies for Sustainable Futures
Promises and Pitfalls
Edited by Chiara Certomà, Fabio Iapaolo, and Federico Martellozzo

Rethinking the Concept of Waste and Mass Consumption
Preserving Resources through Reuse, Repair and Recycling
Richard Waite

Mediating Sustainability in the Consumer Society
Edited by Astrid Skjerven, Lisbeth Løvbak Berg, Liv Merete Nielsen, and Dagny Stuedahl

Smart Cities to Smart Societies
Moving Beyond Technology
Edited by Esmat Zaidan, Imad Antoine Ibrahim, and Elie Azar

Events of Art and Education in Post-climate Times
Carl Anders Säfström and Glenn Loughran

Sustainability Perception Indicators
Theory and Practice of Perception Indicators in Sustainability
Stephen Morse

For more information on this series, please visit: www.routledge.com/Routledge-Studies-in-Sustainability/book-series/RSSTY

SUSTAINABILITY PERCEPTION INDICATORS

Theory and Practice of Perception Indicators in Sustainability

Stephen Morse

Routledge
Taylor & Francis Group
LONDON AND NEW YORK

earthscan
from Routledge

Designed cover image: Dilok Klaisataporn (Getty Images)

First published 2026
by Routledge
4 Park Square, Milton Park, Abingdon, Oxon OX14 4RN

and by Routledge
605 Third Avenue, New York, NY 10158

Routledge is an imprint of the Taylor & Francis Group, an informa business

British Library Cataloguing-in-Publication Data
A catalogue record for this book is available from the British Library

ISBN: 978-1-032-59859-8 (hbk)
ISBN: 978-1-032-59853-6 (pbk)
ISBN: 978-1-003-45660-5 (ebk)

DOI: 10.4324/9781003456605

Typeset in Times New Roman
by Apex CoVantage, LLC

CONTENTS

FIGURES

TABLES

BOXES

FOREWORD

> Joy is the clearest indicator of deep wellness.
>
> (LaPorte, 2014)

I must admit that I like indicators!

Indicators are typically, but not always, numbers designed to help tell us things about the world around us, and the primary reason I am interested in them is because they are created, communicated and used by people. It is the social—the human—side of indicators, no matter how technically sophisticated and complex the maths behind them may be, that I find fascinating. Indicators are a deep reflection of us—our strengths, frailties, wishes and desires—in essence, who we are and who we want to be.

This book is about indicators, but with an unashamed focus on just one type of them—the indicators designed on purpose to capture our perceptions of the world and our place within it. There are many of these indicators, perhaps more than you may think, and they cover all sorts of topics from the state of the environment to economics. Believe it or not, there is even a group of them designed to assess our perceptions of beauty. Just imagine that—reducing something as wonderfully abstract and value-laden as beauty into a set of numbers. Just what would that look like? Would it be something bizarre if not amusing, it may even conjure disturbing parallels with the history of eugenics and the drive to create what some perceive to be a "perfect" appearance.

Yet there are published frameworks based on indicators of beauty and attractiveness, not the same thing of course, for the human form and personality. Indeed, it does not stop with beauty and attractiveness in people; there are also indicators designed to assess our wellbeing and happiness, even though these seem to be so subjective and personal. I will cover all of these in the book.

Indicators based on our perceptions are widespread, and, indeed, we use many of them every day without even realising it. These indicators are not necessarily numbers but clues we get from our senses. A smell of smoke indicating potential danger, the colour of the sky telling us about what weather we may expect and the sound of laughter as an indicator that all is well. Also, perhaps as Danielle LaPorte has said in the quote above, joy is the clearest indicator of a deep wellness even if these are not expressed as numbers.

The particular focus of my exploration into the world of perception-based indicators is sustainability, which I will consider here as being synonymous with sustainable development. I will define what I mean by all of this in a later chapter, but here, it only needs to be said that achieving sustainability has been an ongoing and monumental challenge for humanity. We face so many threats that impact on our wellbeing, let alone the wellbeing of our children and their children, that sustainability sometimes feels like the pot of gold at the end of a rainbow; we keep trying to get there, but are we really making any progress? Indicators represent one of the main approaches we have adopted to help us assess progress, or lack of it, towards sustainability, and for the most part these are technical-sounding measures, such as gross domestic product (GDP), Gini coefficients of income equality, SOx and NOx concentrations in the atmosphere and greenhouse gas emissions (assessed as CO_2 equivalents). Based on their names, these have a very technical, maybe even "scientific" feel to them. But there is a role here for perception, not least because people's perceptions of the world and its challenges help drive our behaviours, and it is our behaviour that matters when it comes to the attainment of sustainability. While it is, of course, vital to measure greenhouse gas emissions, as these drive climate change, the emissions are mostly coming from the burning of fossil fuels (coal, gas, oil) as a source of energy, and we are the ones who do this. Controlling greenhouse gas emissions requires us in large part to burn less fossil fuel; it requires us to change our behaviour. But how do you encourage people to make that change? How is this process of change influenced by attitudes and perceptions? Indeed, how do we even assess attitudes and perceptions? Cue the need for perception-based indicators.

I hope the reader finds this foray into the world of perception-based indicators of sustainability enlightening and intriguing, but above all stimulating, and I also hope it provides much food for thought.

Reference

LaPorte D (2014). *The Desire Map: A Guide to Creating Goals with Soul.* Sounds True Publishing, Louisville, CO

ACKNOWLEDGEMENTS

I would like to thank all those who encouraged and supported me during the time it took to research and write this book, especially my family—wife Maura and children Llewellyn and Rhianna.

ABBREVIATIONS

ACBPI	ASEAN Competition Business Perception Index
ANOVA	Analysis of Variance
ANCOVA	Analysis of Covariance
ASEAN	Association of Southeast Asian Nations
BITOAC	Beneficiary, Implementors, Transformation, Owners, Assumptions and Constraints
CATWOE	Customer, Actor, Transformation, Worldview, Owner and Environment
CEO	Chief Executive Officer
COE	Compensation of Employees
CPI	Corruption Perception Index
CREAM	Clear, Relevant, Economic, Adequate and Monitorable
DF (df)	Degrees of Freedom
DPSIR	Driving force-Pressure-State-Impact-Response
EF	Ecological Footprint
EPU	Economic Policy Uncertainty Index
EU	European Union
GDP	Gross Domestic Product
GHG	Greenhouse Gas (e.g., carbon dioxide)
GNH	Gross National Happiness
GNHI	Gross National Happiness Index
GNI	Gross National Income
GNP	Gross National Product
GOS	Gross Operating Surplus
GPI	Genuine Progress Indicator
GPP	General Population Poll
GRR	Global Risks Report

GWP	Gallup World Poll
HDI	Human Development Index
HPI	Happy Planet Index
IAEG-SDGs	Inter-Agency and Expert Group on SDG Indicators
IAP2	International Association of Public Participation
IEP	Institute for Economics and Peace
IQR	Interquartile Range
ISEW	Index of Sustainable Economic Welfare
Log	Logarithm
LN	"Natural" Logarithm (logarithm to the base e)
MBTI	Myers-Briggs Type Indicator
MDG	Millennium Development Goals
MS	Mean square (found by dividing sums of squares by degrees of freedom)
NEF	New Economics Foundation
NGO	Non-Governmental Organisation
NPS	Net Promoter Score
OECD	Organisation for Economic Cooperation and Development
OPHI	Oxford Poverty and Human Development Initiative
OPN	Opinions and Lifestyle Survey
P	Probability (used here in terms of statistical significance)
PAR	Participatory Action Research
PLA	Participatory Learning and Action
PRA	Participatory Rural Appraisal
PSIR	Pressure-State-Impact-Response
QRQ	Qualified Respondents' Questionnaire
r	Correlation coefficient
R2	Coefficient of determination
RP	Rich Picture
RRA	Rapid Rural Appraisal
SDG	Sustainable Development Goal
SI	Sustainability Indicator
SMART	Specific, Measurable, Achievable, Relevant, Time-bound
SPI	Safety Perceptions Index and also Social Progress Imperative
SPICED	Subjective, Participatory, Interpreted and communicable, Cross-checked and compared, Empowering, Diverse and disaggregated
SS	Sum of Squares
SSM	Soft Systems Methodology
TI	Transparency International
T&V	Training and Visit System
UCM	Unobserved Components Model (also known as "structural time series models")
UK	United Kingdom

UN	United Nations
UNDP	United Nations Development Programme
UPI	Uncertainty Perception Indicator
VIF	Variance Inflation Factor
WCED	World Commission on Environment and Development
WGI	Worldwide Governance Indicators
WHR	World Happiness Report
WISE	Well-being, Inclusion, Sustainability and Equal Opportunity
WJP	World Justice Project

1

INTRODUCTION

Sustainability, indicators and perception

1.1 Relativism and indicators

I would like to begin the book with a couple of short quotations:

> Humans see what they want to see.
>
> *Rick Riordan,* The Lightning Thief

> The mice think they are right, but my cat eats them anyways. This is the point, reality is nothing; perception is everything.
>
> *Terry Goodkind, American writer*

> In politics, perception is reality.
>
> *Lee Atwater, US Republican Party political consultant and strategist, Time, October 3, 1982*

> It is what men think that determines how they act.
>
> *John Stuart Mill, philosopher*

While all the quotes are short, there is a lot to unpack. Do people really see what they want to see? Is our knowledge of the universe, much of which has been gained through science, really driven by what we want to see? Is perception really everything and, perhaps most contentious of all, does perception create reality? I can readily imagine that not all those reading this book will necessarily agree with what is being expressed in these quotes, but like all popular sayings, there is always a grain of truth in what is being said. Our views of the complex interactions we see in the world will be framed at least in part by our beliefs, culture, likes, dislikes,

DOI: 10.4324/9781003456605-1

wishes and wants. We can seek to capture all of that in a set of measurements that we like to think are unbiased, objective and capture "reality", but even here, it has been our choice as to what to measure, how to do the measuring and how to interpret what emerges. Change any of these decisions and the world could look very different. Indeed, there is a whole school of thought called "relativism" which:

> is the view that truth and falsity, right and wrong, standards of reasoning, and procedures of justification are products of differing conventions and frameworks of assessment and that their authority is confined to the context giving rise to them.
>
> *(Stanford Encyclopedia of Philosophy)*

Or, put more succinctly, relativism claims that "knowledge is relative to the limited nature of the mind and the conditions of knowing" (*Merriam-Webster Dictionary*); the knowledge we have is influenced by what we want to know and how we want to know it. If this is so, then there is no true objectivity, as our attempts to seek knowledge will be influenced by our very humanity—what it means to be human with our accumulated beliefs, culture and so on.

Relativism certainly has its critics, and many have argued that there are, for example, laws of nature which are discoverable and hold no matter what the beliefs or biases of their discoverers. If relativism were true, then how is it we were able to discover the laws of nature and apply that knowledge to generate the goods and services that we use every day? Indeed, as one critic of relativism has succinctly put it:

> I do not want to drive across a bridge designed by an engineer who believed the numbers in structural stress models are relative truths.
>
> *(R.C. Sproul,* Choosing My Religion*)*

Indeed, relativism is at odds with itself in its fundamental claim that there is no such thing as truth. If all knowledge is relative, then relativism is itself relative and thus cannot be true!

> *Fundamentally it consists in propounding the claim that there is no truth as if this were truth or in declaring it to be absolutely true that there is nothing but the relatively true; one might just as well say that there is no language or write that there is no writing.*
>
> *(Frithjof Schuon,* Logic and Transcendence*)*

I do not intend going any deeper into this philosophical debate of relativism versus science, as the brain can start spinning, but I would argue that there is one realm where relativism does have an undoubted sway—the world of indicators.

Indicators are a highly diverse family of tools that can be found in many areas of interest to us, from biology (e.g., indicators of biodiversity) to economics (e.g., indicators of economic performance) and society (e.g., indicators of wellbeing). Indeed, it sometimes seems as if they are everywhere, and you may not realise it but you are probably confronted with many indicators each day, and indeed, you probably use them yourself. But what are these ubiquitous indicators? How do I know them when I see them? There are numerous definitions of an indicator, and the following are just two of the shortest I can find, both taken from dictionaries. I am not claiming these definitions are necessarily the best available, but they are certainly short and to the point:

> A thing that indicates the state or level of something.
>
> *(Oxford Dictionary)*

> An indicator is a measurement or value which gives you an idea of what something is like.
>
> *(Collins Dictionary)*

The first of these is not especially enlightening; an indicator is something that indicates something. In the same vein of thinking, a hammer can be defined as something that can be used to hammer things! I include it here as an illustration of the circularity we can easily get into when trying to define something. But the second definition makes use of the term "measurement" and that takes us into a different territory. Some people often use the term "metrics" as being synonymous with indicators, and this is understandable given that the *Oxford Dictionary* defines metrics as:

> A set of figures or statistics that measure results.
>
> *(Oxford Dictionary)*

But the difference is that indicators are designed to provide an idea of a state or level of something so as to help provide a guide. A simple example here would be the heating of a metal rod. As we heat and cool the rod, we can measure its characteristics, such as temperature and expansion/contraction, as precisely as we wish by using a variety of instruments available to us, and those temperatures will provide us with a table of metrics—a set of numbers showing how the metal expanded and contracted depending on temperature. These metrics can tell us a lot about how different metals, including perhaps mixtures (amalgams) of metals, behave when heated, and that can be very important in various industrial processes. But even without the use of these instruments to measure temperature and expansion precisely, we can get a rough idea of the temperature of the rod. We can look at its colour—hot metal tends to become red, then eventually white—place drops of water on its surface—as the metal heats, the water drops will fizzle and evaporate very fast—or, and this one is certainly not recommended, we could touch the surface with a finger. These signs (colour, fizzling of water, pain and smell from the burning of fingers) do not tell us the exact temperature of the rod in the same way that instruments can, but they provide a guide; they are indicators of the temperature of the rod. The major advantage with these indicators, although the use of pain as an indicator has an obvious disadvantage that outweighs any advantages, is that we do not need to have expensive measuring equipment to tell us the exact temperature, and maybe only a rough idea is all that is required for those who work with such metals. Maybe the latter do not need precise metrics of temperature for their day-to-day work with metals, and indicators, along with experience of course, would be enough.

The change of colour, the fizzing of water and even the burning of a finger when metal is heated are all based on our senses of seeing, hearing and touch. This brings us to an often overlooked point that indicators can be both quantitative (i.e., numerical) and, perhaps, look and feel like metrics or qualitative (i.e., more subjective). There are obvious overlaps between this binary division into quantitative and qualitative. After all, it is possible to "score" qualitative attributes such as colour. An experienced metalworker could be asked to score the colour of heated metal from grey to red, orange, yellow and eventually white, and indeed, there are published scales that show the rough link between colour and temperature. This conversion of something that is subjective into numbers will be returned to later in the book, but it is fair to say that for the most part, indicators have been conceptualised and realised far more in quantitative terms than qualitative, even though the latter format is arguably the one we use most in our daily lives.

Indicators are designed to provide us with windows into the complexity of the world in which we live. But which indicators to choose, how to give them values (i.e., populate them) and how to interpret what those indicators are telling us are all based on human judgement. Those choices are indeed framed by what we think is important to understand, represent and, most critically of all, act upon. They are akin to the windows of the New York Cityscape in Figure 1.1. The entire cityscape is a complex and multilayered and faceted mix of different buildings in terms of style and age but also use. We can all see this cityscape, but each of us will have our own interests and focus, as epitomised by the four windows in Figure 1.1. Some of us will see the Statue of Liberty as the highlight and bring with it feelings of welcome and freedom. Others will focus on the older buildings and their representation of history and tradition. For others, the focus will be upon buildings such as City Hall and its association with democracy, governance and politics, while for others, it will be buildings in the financial district and their association with money and power. For others, and I include myself in this group, the take-away message may perhaps be one of wonder, awe and majesty at the sheer scale of the buildings and how they have been built; the way in which they seem to reach to the sky. But the key message here is that each of us chooses the window(s) we wish to use, and those decisions will be framed in part by who we are and the numerous influences that have helped shape our lives and continue to do so. There are many such windows, many more than the four in Figure 1.1, and while the views may be

FIGURE 1.1 Different perceptions of the same urban landscape

Diagram comprises four windows on the same cityscape landscape, and the words that may be associated with those views.

different, each of them is equally valid; there is no such thing as "better" or "worse" windows here.

Indicators are also ways of seeing the world but with the important difference that the windows we look through have been selected and designed by others and presented to us with a purpose. Indeed, indicators are not designed to be passive; their very existence is driven by a desire to help bring about a desired change. But these decisions surrounding choices of indicators and how best, if at all, to make use of them to help inform decisions are made by people; there are no laws of nature which apply here. In that way, we can think of indicators as being children of relativism; they are what we choose them to be, and those choices will be framed by what we value and how we wish to see the world. They are not objective truths arising from a law of nature but represent the wishes of their creators. It is easy to appreciate how we may lose sight of this. Presenting indicators as numbers, perhaps with elaborate methodologies, including the use of sophisticated instruments and data manipulation using statistical techniques, may generate an aura of objectivity and scientific precision, and this can especially be so if those presenting the indicators are from a well-respected organisation. But there is no getting away from the reality that indicators are the windows that someone has selected for us to see through, and there is a purpose behind these decisions.

The focus in this book is upon one type of indicator—those designed to help us engage with the highly complex topic of sustainability. Much has been written about sustainability indicators (SIs) (Morse, 2019), but here the emphasis will be upon the development, assessment and use of these indicators based on perceptions.

The aim is to set these out for the reader along with the issues associated with them. The reader may see at various points that the term "index" (plural indices) is used instead of indicator, and these are different; an index is an amalgam of separate indicators and can be quite complex in terms of its structure and methodology. In some cases, an index may comprise literally dozens of indicators with all kinds of different weightings and means of integration. To help simplify mattes, the term "indicator" will be used throughout the book rather than the more generic and cumbersome but technically correct term "indicators and indices". Using "indicator" as a generic term to cover indicators and indices is, admittedly, something of an uncomfortable fudge, and no doubt will attract some criticism but hopefully it helps avoid the use of too many acronyms and does help make the text more digestible.

1.2 Perception: A mental image

Perceptions are the basis of our interaction with the world around us, and our perceptions are influenced by many things including our experience, personality, culture and what other people tell us. Turning to dictionaries once again, we learn that perceptions are:

> a thought, belief, or opinion, often held by many people and based on appearances.
>
> *(Cambridge Dictionary)*

> a mental image.
>
> *(Merriam-Webster Dictionary)*

> Your perception of something is the way that you think about it or the impression you have of it.
>
> *(Collins Dictionary)*

> the way you think about or understand someone or something.
>
> *(Britannica Dictionary)*

All these definitions are rather nebulous and do seem to have overlaps with other terms, some of which are embedded within the definitions such as opinion, belief, understanding and impression. Our perceptions of the world are framed by many factors, including our culture, gender, upbringing and biases, and on a day-by-day basis, we are often highly aware of our surroundings and the signals we see that point to dangers and opportunities. We use these signals all the time, including in our interactions with others. A smile, a frown, a stare—all of these and many other body language clues, as well as language are used by us every day to "indicate" our circumstances, provide a mental image and how to respond. There is also inevitably a large degree of variation between us in terms of our perceptions, but there may also be some commonalities. There are biases, conscious and unconscious, at play here and much can depend on experience; framing expectations from the signals, we see based on previous experience. Indeed, our perception of the world around us is based on many clues and will also be framed in part by our beliefs and opinions as noted in the definition above from the *Cambridge Dictionary*, and personality can also play a role. Indeed, given that perception is partly framed by our beliefs, then we can arrive at very different perceptions of the same thing, be it people, events, ideas, objects or whatever.

Judgement is a term often associated with perception and describes the process of forming an opinion or understanding as well as coming to a decision:

> the process of forming an opinion or evaluation by discerning and comparing.
>
> *(Merriam-Webster Dictionary)*

> the act or process of forming an opinion or making a decision after careful thought.
>
> *(Britannica Dictionary)*

In both these definitions, the emphasis is on judgement being based on careful thought—discerning and comparing—although these "careful thoughts" will also be framed by perception. Hence, both perception and judgement are mental processes inherent within all of us and are reflections of personality, attitudes, values, beliefs and so on. Understanding perception and judgement is thus a challenging subject area for research, but even so, they are of much interest and have attracted much attention. In business, for example, the perception of existing or potential consumers of products and services is often sought as a means of making changes to improve marketability. For example, one theoretical framework often utilised as the basis for research in perception is based on the concept of dissatisfiers and satisfiers (Cadotte and Turgeon, 1988). Dissatisfiers are those who expect things to be in good order but will make a negative judgement if it is not. On the other hand, if the thing is in good order, they do not pass a positive judgement as they expect it to be good and will express dissatisfaction if it is not. Satisfiers are those who do not have a default expectation of good order but will express a positive judgement if they experience something they like. For online marketing where customers can readily express their views via feedback, and the "ratings" that these generate can influence decisions of other consumers, it is easy to see why there has been so much interest in this topic.

As personality has long been known to be one of the factors that influences both perception and judgement (what we decide to do about what we perceive), then that too has been the subject of analysis, and an example is the Myers-Briggs framework for understanding the types of personality (Myers and Myers, 1980; McCrae and Costa, 1989). The Myers-Briggs framework (Table 1.1) comprises a

TABLE 1.1 The Myers-Briggs framework for understanding the types of personality. After Myers and Myers (1980) and McCrae and Costa (1989).

Theme number	*Theme*	*Example question*	*Preferences*
1	Opposite ways to direct and receive energy	*Do you prefer to focus on the outer world or your own inner world?*	Extraversion (E): outer world Introversion (I): inner world
2	Opposite ways to take in information	*Do you prefer to focus on the facts or the big picture?*	Sensing (S): facts Intuition (N): big picture
3	Opposite ways to decide and come to conclusions	*Do you prefer to take an objective or an empathetic approach for deciding?*	Thinking (T): objective Feeling (F): empathetic
4	Opposite ways to approach the outside world	*Do you prefer to seek closure or stay open to new information?*	Judging (J): seek closure Perceiving (P): stay open

set of eight preferences set out as four pairs of opposites. A combination of these preferences (E or I, S or N, T or F, J or P) can be used to create a Myers-Briggs Type Indicator (MBTI) comprising a four-letter code (e.g., ESTJ) which is an assessment of your psychological or personality type. There are 16 possible combinations of the preferences which results in 16 distinct MBTI personality types capturing differences in interests, views and behaviour.

One of the four themes in the Myers-Briggs framework is titled "Opposite ways to approach the outside world" with the associated question being "Do you prefer to seek closure or stay open to new information?" Those who seek closure are assessed as "J" (Judging), while those who stay open-minded are assessed as "P" (Perceiving). It may look relatively simplistic—with just 16 possible personality types, and it certainly has its critics (McCrae and Costa, 1989), but it is still widely used and does have solid roots in the work of the Swiss psychiatrist Carl Gustav Jung. Jung argued that variation in behaviour between people may appear to be random, but this can be explained by differences between them in the ways they apply perception and judgement. The Myers-Briggs framework is regarded by some as being robust, as noted on the Myers and Briggs Foundation website (www.myers-briggs.org/):

> There has been more than 70 years of research on the MBTI system. The MBTI instrument is scientifically reliable and valid.

The framework can help "people identify and gain some understanding around how they take in information and make decisions, the patterns of perception and judgment, as seen in normal, healthy behavior". Indeed, as Isabel Briggs Myers, the co-creator of the framework along with her mother (Katherine Cook Briggs), has said, with emphasis added by me (source: www.myersbriggs.org/):

> Whatever the circumstances of your life, the understanding of type can make your *perceptions clearer*, your *judgments sounder*, and your *life closer to your heart's desire*.

That is quite a promise.

1.3 Perception-based indicators

Indicators based on gauging people's perceptions of something are not in themselves a new concept; they have been around for a very long time. Indeed, as noted above, such indicators can include the clues we see every day in the world around us and which we use to guide our actions. These can be regarded as "informal" indicators—indicators that do not necessarily have a numerical value and are not part of any published framework intended to provide guidance. That is not to say, of course, they cannot have a value, as shown with the earlier example of scoring colour change when metal is heated. Similarly, we often look to the sky to see if it is clear and blue or cloudy and grey as an indicator of what the weather is and may be; a grey, overcast sky is an "indicator", which suggests we need to take an umbrella with us when venturing outdoors. We do not need to measure the degree of cloudiness to make use of such a visual "indicator"; we do it all the time without consciously thinking about it, but there are published scales for assessing the level of cloudiness, and one of these is called the Okta (Figure 1.2). The Okta is an eight-point scale ranging from 0 Oktas, which is a clear blue sky, one Okta being 1/8 of the sky covered in cloud, and so on up to 8 Oktas, which represents a sky that is completely overcast. Most of us do not use the Okta as we have an instinctive "feel" for looking at the sky as a way of thinking whether it may rain or not, but the formal scale does exist and is widely used by meteorologists. Even so, while the Okta is based on something that can be measured using instruments, perhaps via satellites, it is also amenable to assessment via perception; the opinion we may have of the degree of cloud coverage.

This potential to provide a numerical value to almost anything we perceive makes it challenging to define what a perception-based indicator is beyond the circular statement that such an indicator is based on gauging perception! A value for the Okta scale can be provided by simply looking at the sky and estimating the extent of cloud cover, but it can also be measured, presumably quite accurately, using ground-based instruments looking up or via aircraft and spacecraft flying above the clouds looking down. An Okta value generated by a perception of cloud cover is thus a perception-based indicator, while measurement of cloud cover via an instrument is a non-perception-based indicator. There are clearly overlaps here, and while the means of giving values to the Okta scale (i.e., "populating" the scale)

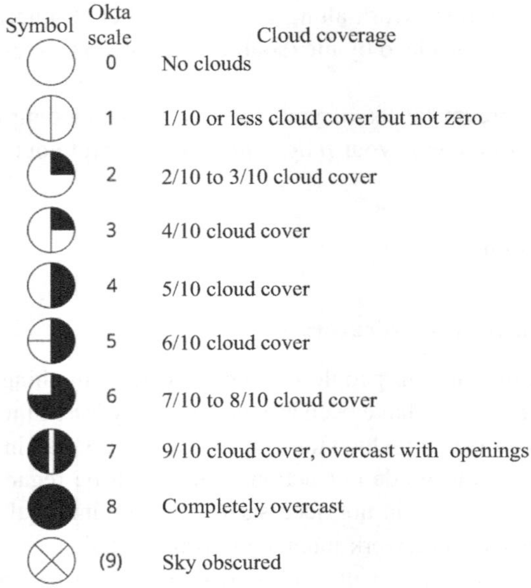

FIGURE 1.2 The Okta scale of cloud cover, along with the meteorological symbol for each point on the scale

Source: Adapted in part from a diagram licensed under the Creative Commons Attribution-Share Alike 3.0 Unported license.

are different, we may expect results to be related although not necessarily in a linear way. Maybe we can judge the extremes of the Okta scale quite well but struggle with those towards the middle.

Another way of thinking about perception-based indicators is to consider the context of their use. These indicators are especially useful in contexts where it can be challenging, for all sorts of reasons, to measure what is really happening. The challenge may simply be one of cost. In the Okta example above, the use of instruments on board satellites or aircraft to measure cloud cover may be expensive, not just in terms of the instrument themselves, although that can be significant, but the cost of technicians able to operate the machines, analyse and interpret the data. It is obviously far cheaper to simply look at the sky and come up with an estimate of cloud cover. But there are other contexts where the challenge is much greater than just cost, and perhaps, a classic example of this is with the assessment of crime (Lohr, 2019)—a topic that often attracts much public and indeed political interest. There are many indicators designed to assess crime and its impacts, but one of the most commonly cited is the crime rate defined as the number of crimes, maybe sub-divided into the type of crime (e.g., burglary, violent crime) that take place per unit of population (e.g., number of crimes per 1,000 residents) in a defined period of time (e.g., 1 year). It is obviously an important metric, but how can it best be assessed? This is not as straightforward as it may first appear. An obvious starting point would be to measure the number of reported incidents of criminal activity, but the keyword here is "reported". For the most part (there are no doubt some exceptions), criminals do not prioritise the formal reporting of the crimes they commit; far from it. If criminals could somehow be encouraged to record and report their crimes, then that would be the best possible data for assessing crime rates, but unfortunately this is likely to be highly unreliable to say the least!

An alternative approach to the self-reporting of crimes by those who commit them is to utilise reported incidents made by the victims of crime to the police and other authorities. Unfortunately, there are reasons why people do not formally report crimes, including fear of the police and a lack of faith and trust in the ability of the police to bring the criminals to justice. Indeed, governments often run anonymised surveys, referred to be by some as "victimization surveys" (Lohr, 2019), to determine people's experience of crime and the impacts it has had on them, and these typically generate higher rates than relying on recorded incidents of crime. But can people's claims of being a victim of crime be trusted? Indeed, going beyond the assessment of crime rate, people can also be asked about their perception about the chances of becoming a victim of crime. Here, they are not being asked to recount personal incidents of being a victim but, in essence, to provide a perception of the level of crime in their community. However, the latter is likely to be influenced by a host of factors, including media reports and conversations with friends and family. While perceptions of crime are easy to dismiss given all this influence and potential for bias, they are nonetheless a significant driver of behaviour.

All the above approaches will generate different "signals" as to the crime rate (Lohr, 2019) and that can matter when it comes to decisions made by the police and indeed others, such as politicians, as to the allocation of resources and indeed prioritisation. But such decisions over what crimes to prioritise by the police may also be influenced by perception, and opinions can be very divided. In the UK, for example, the content of social media can potentially be the basis for a criminal offense and thus subject to investigation by the police. But where to draw the line between what some may see as "hate" being spread via social media while others may see it as nothing more than a robust yet valid criticism within freedom of speech and thus a fundamental right? To illustrate this dichotomy, here are some very different takes on the power of social media. The first quotation reports the views of Sir Nick Clegg who used to be a senior politician in the UK but left politics to become the vice-president of global affairs and communications at Facebook—a global social media platform. Perhaps unsurprisingly given his position, he is critical of what he regards as so-called "speech crimes":

> Sir Nick Clegg has said British police have become "too censorious" of online speech. The former deputy prime minister said forces had adopted "whack-a-mole" approach to policing the internet. Speaking at the Charleston Literary Festival in Sussex on Thursday, Sir Nick said "so-called online speech

> offences" were often just about remarks "that people might say to each other in a pub". . . . Hundreds of people have been charged with online "speech crimes" in the past two years. Almost 300 people have been charged with spreading illegal "fake news" or sending "threatening communications" since the Online Safety Act came into force in 2023. Dozens have received convictions under the act.
>
> *(Leake, 2025)*

The ramifications of such "speech crimes" have also been highlighted by some in the media and here is an excerpt from an article published in the Daily Mail newspaper in the UK:

> A former PC [Police Constable] tweeting a joke about identifying as a fish and a boy calling another a "leprechaun" are further examples of so-called "hate incidents" investigated by police. Non-crime hate incidents (NCHIs) are meant to be reserved for cases that are "clearly motivated by intentional hostility" and where there is a genuine risk of significant escalation, according to government guidance. But the police have been accused of wasting "valuable time" on investigating jokes and playground insults as possible hate speech amid a rise in serious offences like knife crime.
>
> *(Tingle, 2024)*

But views do vary on this matter, and some have pointed to the potential harm that such online content can create. Here is a quote from the US, which picks up on a decision by Meta, the owner of Facebook, to remove fact-checkers from its system and thus facilitate "unfettered fee speech".

> Policing the truth on social media is a Sisyphean challenge. The volume of content—billions of posts in hundreds of languages—makes it impossible for the platforms to identify all the errors or lies that people post, let alone to remove them. Yesterday, Meta—the owner of Facebook, Instagram and Threads—effectively stopped trying. The company said independent fact-checkers would no longer police content on its sites. . . . Elon Musk purchased Twitter (which is now called X) with a promise of unfettered free speech. He also invited back users banned for bad behavior. . . . Numerous studies have shown the proliferation of hateful, tendentious content on X. Antisemitic, racist and misogynistic posts there rose sharply after Musk's takeover, as did disinformation about climate change.
>
> *(Myers, 2025)*

The rights or wrongs of these disparate views and where any line should be drawn between free and hate speech are not my focus here; these are very deep waters, and the reader can come to their own conclusions. Banaji and Bhat R (2021), for example, provide case studies of the relationship between social media and hate speech spanning a range of geographical contexts including Brazil, Myanmar, India and the UK. However, the key point here is that there are multiple perceptions at play, even if each of them feels that their own perception is the correct one and supported by what they regard as evidence.

While the use of indicators to capture a community's perception of crime may be useful, and even those in favour of more "objective" reporting may concede that it is still useful to appreciate what the community "perceives" crime levels to be, there may still be perceptions at the heart of even the most (apparently) objective indicators. There is relativism here as these decisions over what indicators to select, how to measure and use them are all driven by people, and different people may make different choices. While some may see this as a simple dichotomy into two broad types of indicator—those that set out at the onset to measure perception (i.e., perception-based indicators) and those that are intended to measure "reality" (i.e., reality-based indicators)—I would argue that both are subject to relativism. Even indicators that we like to think of as "objective", "scientific" and measuring a "reality" may not be as "objective", "scientific" and "reality" focussed as we may think or indeed wish them to be. However, and critically, this is not in any way meant to demean or downplay the value of indicators, be they based on perception or not. They do have great value in helping to capture and convey complexity to people who are meant to act on information and, as we will see later, to help raise awareness and to facilitate debate and discussion. But care does need to be taken. At the beginning of this chapter, I included a quotation from a critic of the reach of relativism and in particular its extension into science:

> I do not want to drive across a bridge designed by an engineer who believed the numbers in structural stress models are relative truths.
>
> *(Sproul, 2005)*

This is, of course, a very fair point, and it is hard to disagree when put this way. Equally, I would not want to cross a bridge which had been designed and built based on a few selected indicators (e.g., how appealing the bridge is to look at) that may have little if anything to do with its structural integrity and tolerance to stress.

1.4 Aim, structure and approach taken in the book

The aim of this book is to explore the world of perception-based indicators and how they relate to sustainability. Much has been written about SIs, and it is simply not possible, or indeed desirable, to try and cover all that material here, and the interested reader is referred to texts such as Morse (2019). Instead, the focus of the book will be upon that subset of SIs that are founded on assessments of perception and how they play a role in the complex topic of assessing sustainability. The book has been written as a text for students, primarily those at university level (undergraduate and postgraduate) undertaking degrees in a wide range of disciplines—sustainability is, after all, highly multi- and inter-disciplinary. Hence, students taking programmes in the social sciences, economics, environmental studies, human geography, politics, business studies and international development will hopefully find the material covered here of much interest.

The book has a total of six chapters, including the introduction. Chapter 2 sets out a brief outline of sustainability and how its meaning and attainment can be subject to perception. Included in this chapter is an outline of SIs, along with examples taken from the Sustainable Development Goals (SDGs) framework set up by the United Nations that is the dominant "game in town" when it comes to our efforts to improve the health of the planet and our development. Chapter 3 focusses specifically on examples of perception-based indicators of relevance to sustainability and the key issues that emerge from their development and application. There are many such examples that could have been used, but space only allows a relative few of them to be considered in any depth. The ones selected were not chosen because I think they are necessarily the best, or for that matter, the worst, such indicators, but instead they represent a span of concerns within sustainability and raise issues in terms of how they are designed and given values (populated). If the reader is upset because I have left out their own favourite indicator, then I do apologise, but there are always limits to what can be covered in a book of this length. Chapter 4 explores the participatory approach to the development and use of SIs and how SIs based on perception emerge from this. Chapter 4 brings us back to the relativism already covered at some length in this chapter but with some quite different contexts. Chapter 5 looks at the use of SIs, although in this case that is limited to their use as communication tools within reporting via the traditional media (i.e., newspapers). The question asked in Chapter 5 is whether perception-based SIs have any advantage when it comes to media reporting over those SIs that are not designed to assess perception? Also, how does newspaper reporting address the perception-based nature of these SIs? The final chapter of the book (Chapter 6) looks at the future of perception-based SIs and presents a typology of indicators as part of that exploration.

My focus in the book, as stated at the start of the foreword, is much more on the human side of these indicators rather than the mathematics behind them, but there is an important relationship at play here; the maths behind all indicators is, after all, nothing more than an expression of human choice. Hence, I have avoided a more formal "monograph" style, although I have included references to help support key points and as prompts for further reading, especially when it comes to the questions I have included at the end of each chapter. These questions are certainly not intended as formal essay or exam topics. Instead, they are meant to be prompts ("food for thought") to encourage an exploration of points raised in the chapter. The references included with each chapter will help as a starting point for addressing the questions, but the reader is encouraged to go further and source other material. This is why I have opted to include references at the end of the chapters rather than just a single listing at the end of the book.

I do need to add a few words here about the use of acronyms. Unfortunately, the indicator "multiverse", that multi-dimensional space that encompasses all aspects of indicators from conceptualisation, development, population with data and communication through to use is replete with the use of acronyms. They seem to have grown over the decades, and these days it is almost impossible to come across any publication on indicators that is not full to the brim with acronyms. Indeed, there are now many overlaps where the same acronym can cover different indicators, or frameworks or whatever. CPI, for example, can mean Corruption Perception Indicator or the Corporate Perception Indicator. This can make for a bewildering read backed up with lots of cross-checking to follow what is being referred to. But at the same time, the avoidance of acronyms can generate rather dense text. Hence, I have tried to strike a balance by limiting the use of acronyms as much as possible, and those I have used repeatedly are provided as a list at the start of the book.

Finally, I had something of a decision to make as to whether to call the focus of the book "perception-based indicators", "perceptual indicators" or, against my better judgement, use an acronym (PBI or PB). In line with my desire to avoid acronyms as much as possible, I decided to avoid PBI or PB. I have instead opted for "perception-based indicators", as I have started to do in this chapter rather than "perceptual indicators" largely because I feel the former is a better expression of the nature of the tools at the heart of the book.

1.5 Summary

SIs are useful tools for summarising and presenting key information and what can be highly complex systems. Interventions need to be evidence-based to have the best chances of delivering successful outcomes, and indicators are one way in which information can be provided to help provide guidance to those who need to design, implement and monitor interventions. But all indicators are human constructs; they are founded on decisions made by people as to what to measure, how to present the results and to whom with what purpose. As indicators are human constructs with a purpose that has been perceived and contextualised by their creators, then the concept of relativism becomes relevant. Relativism is a concept that assumes the knowledge we have and encapsulated in indicators is influenced by what we want to know (i.e., the indicators we feel we need) and how we want to know it (i.e., the ways in which the indicators are populated and communicated).

Indicators can be populated with data derived in many ways. In some cases, for example with indicators designed to convey information on environmental pollution, the data may come from scientific instruments, while in others, the data may come from surveys which, in some cases, ask for people's views and opinions. Indeed, a SI framework may comprise a mix of these types. However, even the apparently objective and scientific indicators, populated with data derived from measuring devices, are still the constructs of those who create and promote them. Change the suite of indicators or the ways in which they are populated and portrayed, and the messaging that comes out of this may be significantly different, and this in turn may influence the design and implementation of interventions.

This book focuses on one type of indicator designed to capture and present the perceptions of people about the world around them. While much has been written about the landscape of SIs, the topic of perception-based indicators and how they can play a role in this landscape has received very little attention. Capturing perceptions with indicators has its advantages, especially when it comes to issues that can be challenging to assess in a direct sense, and in this chapter, the example used to convey this point was that of crime. There are various ways of assessing the rate of crime in a community, and they all have their own advantages and disadvantages. One of these is to assess people's perceptions of the crime rate and that is important as it may well influence how people behave. These assessments of crime rate are, of course, influenced by how different crimes should be prioritised by law enforcement agencies, and this can be highly contentious as illustrated with the example of free speech in social media.

1.6 Food for thought

1. What are the similarities and differences between perception and perspective?
2. Complete the Myers and Briggs framework based on your personality (www.myersbriggs.org/). What did you think of the assessment and what are the strengths and weaknesses of this approach?
3. Besides the crime rate, what other challenging topics do you think would be amenable to the use of perception-based indicators?
4. The chapter provides an illustration, in this case, focussing on free speech in social media of how people's perceptions can help influence their views on the prioritisation of different types of crime. Is it possible to reconcile such disparate views and, if so, then how?

References

Banaji S and Bhat R (2021). *Social Media and Hate. Routledge Focus on Communication and Society*. Routledge, London and New York

Britannica Dictionary. Available at: www.britannica.com/dictionary

Cadotte ER and Turgeon N (1988). Dissatisfiers and Satisfiers: Suggestions from Consumer Complaints and Compliments. *The Journal of Consumer Satisfaction, Dissatisfaction and Complaining Behavior* 1, 74–79

Collins Online Dictionary. Available at: www.collinsdictionary.com/

Leake N (2025). Britain's Police Have become 'Too Censorious', Warns Nick Clegg. *The Telegraph*. Available at: www.telegraph.co.uk/news/2025/05/22/uk-police-have-become-too-censorious-warns-nick-clegg/?msockid=2041d0e17d5a62a70273c49e7cba631c

Lohr L (2019). *Measuring Crime: Behind the Statistics. SA-CRC Series on Statistical Reasoning in Science and Society*. Chapman and Hall/CRC

McCrae R and Costa P (1989). Reinterpreting the Myers-Briggs Type Indicators from the Perspective of the Five-Factor Model of Personality. *Journal of Personality* 57, 17–40

Merriam-Webster Dictionary. Available at: www.merriam-webster.com/

Morse S (2019). *The Rise and Rise of Indicators: Their History and Geography*. Routledge, London
Myers and Briggs Foundation. Available at: www.myersbriggs.org/
Myers IB and Myers PB (1980). *Gifts Differing: Understanding Personality Type*. Davies-Black Publishing
Myers SL (2025). Wildfires in Southern California. Plus, We Explain Why Facebook Is Getting Rid of Fact-Checking. *New York Times*. Available at: www.nytimes.com/2025/01/08/briefing/southern-california-wildfires.html
Oxford English Dictionary. Available at: www.oed.com/
Riordan R (2006). *The Lightning Thief*. Hyperion Books for Children, New York
Schuon F (1984). *Logic and Transcendence*. HarperCollins Publishers Ltd, Glasgow
Sproul RC (2005). *Choosing My Religion*. Presbyterian and Reformed (P&R), Phillipsburg, NJ
Stanford Encyclopedia of Philosophy. Available at: https://plato.stanford.edu/
Tingle R (2024). The 'Hate Incidents' Probed by Police: From a Tweet about Identifying as a Fish, to a Boy Calling another a 'Leprechaun, How Cops Are 'Wasting Time' on 'Ridiculous' Cases. *Mail Online*. Available at: www.dailymail.co.uk/news/article-14114131/The-hate-incidents-probed-police.html

2

SUSTAINABILITY INDICATORS

2.1 Introduction: Seeing is believing

> Perceptions refers to the way an individual observes, understands, interprets, and evaluates a referent object, action, experience, individual, policy, or outcome.
>
> *(Bennett, 2016; page 585)*

> Perception is a clash of mind and eye, the eye believing what it sees, the mind seeing what it believes.
>
> *(a quote ascribed to Robert Breault)*

Before we can address the meaning of the term "sustainability indicators" (SIs), we first need to set out what we mean by sustainability, and this is a significant challenge. There are many definitions of sustainability and sustainable development, and some do regard these as being quite different, found in all kinds of publications and websites. But to begin with, it is perhaps easier to illustrate what we mean by unsustainability given that we have, sadly, witnessed many examples of this, including many in my lifetime. Hence, the chapter begins by setting out an example of a detrimental environmental impact that is still very much with us and in the news, namely human-induced climate change. Notice that I have said "human induced" climate change resulting from what is colloquially referred to as "global warming"; the global increase in average temperature as a result of the release of "greenhouse" gasses (or GHGs). Much has already been written about this important topic, and I certainly do not intend to go over that same ground, but instead I will focus on the role that public perception plays and how researchers have tried to capture that perception using an indicator.

DOI: 10.4324/9781003456605-2

The second section of the chapter focusses on some definitions of sustainability followed by an outline of SIs as tools to help us with the attainment of sustainability. My focus in the latter will be upon the SDG framework of goals, targets and indicators agreed to by many countries across the globe. The SDGs really are the big "game in town" at present when it comes to sustainability and they matter.

The final section of the chapter will explore perception-based indicators within the SDG framework. These indicators are a subset of that wider group of SIs, so it is first necessary to set out what these are and provide some examples taken from the SDG framework. I will provide many more examples of perception-based indicators in Chapter 3 and discuss issues associated with them in greater depth, but here, my aim is to provide some examples and set out how they are different from other indicators in the SDG framework. This "spectrum" or typology of types of SI will be revisited and built upon in later chapters of the book, especially Chapter 6.

2.2 Public perceptions of environmental challenges

Public perception has often been regarded as an important dimension when it comes to addressing environmental issues. This may perhaps sound a little odd, as surely such issues are the ones we expect to be studied by scientists, and the evidence they come up with is presumably objective and can be used, with confidence, to plan interventions that would succeed. What role can there or indeed should there be for public perception in all of these? Can scientists and others we trust just be left to get on with it. Well, the answer to this seemingly simple assumption becomes more complex when we need to ask people to change their behaviour and/or accept changes imposed by others such as via legislation. After all, while interventions may be based on solid scientific evidence, they may have little traction with the public if they think they will lose out in some way. Even if attempts are made by those with power to communicate, encourage or even force change in behaviour, there is always the possibility, at least in a democracy, that the public will vote them out at the next election if they do not like what they are hearing and seeing. There can, unfortunately, be a tendency for people to agree with the "doing" of something so long as it is someone else who pays the price. Hence, understanding perceptions can be very important when it comes to maximising public support for policy and interventions, and perhaps unsurprisingly given its importance, this is a topic that has been well-researched (Bennett, 2016). As Bennett (2016; page 589) has noted, in this case linked to nature conservation:

> Perceptions provide insight and are an indispensable form of evidence that deserves a central place in the toolkit for monitoring, evaluating, and adapting conservation programs and policies. . . . Perceptions can and should be integrated with other forms of evidence as part of robust monitoring and evaluation protocols. This information can then be communicated and incorporated into deliberative policy making and adaptive management for conservation.

Yet any perceptions of the risks associated with environmental challenges may often be based on limited knowledge amongst the public of these challenges and their impacts:

> Risk perception and risk judgments become more and more important in modern society (Beck, 1992), especially with regard to environmental problems. The nature, seriousness, and the consequences of most environmental problems are uncertain and are unfamiliar to people. Therefore, the way one deals with these environmental problems is to a large extent dependent on risk perception and risk judgment.
>
> *(Steg and Sievers, 2000; page 264)*

After all, we can readily imagine that these environmental problems may differ significantly in terms of who are directly and indirectly impacted. Those who feel they are not likely to be impacted may wish to avoid paying the price (economic, social or otherwise) of interventions designed to avoid or mitigate the impacts arising from damage to the environment; it is not affecting me so why should I pay? However, there may be others who feel they could be negatively impacted by damage to the environment and very keen to see such interventions happen irrespective of the cost, especially if that cost is paid by others! Indeed, it is possible that someone may not even realise they are being negatively impacted by damage to the environment especially if this happens over longer time scales and is accumulative! There is a tangled web of perception here but unpacking all this matters when it comes to the likelihood of success of interventions; after all, changing the behaviour of a few or even a majority may not be enough.

Today, of course, we are in the midst of an environmental impact having global ramifications caused by "global warming". Climate change was not invented by humans and has been a feature of the planet ever since the creation of the atmosphere billions of years ago. But we have witnessed a significant and rapid warming since the industrial revolution in the 19th century, and the evidence available to us strongly suggests the primary cause has been the release of GHGs (such as carbon dioxide, methane and even water vapour) into the atmosphere mostly as a result of the burning of fossil fuels (coal, gas, oil, etc.). These gasses, along with water vapour, create a "greenhouse effect", whereby the sun's radiation is trapped near the planet's surface and results in warming. This warming has a complex set of consequences when it comes to climate and can also lead to the release of more GHGs such as water vapour—what is referred to as "feedback". Nested within all these complex processes are what climate scientists refer to as "tipping points"; when the temperature reaches a certain level, the feedback starts to make things a lot worse a lot quicker. While this is complex, the physics behind such cause–effect are well-understood and based on multiple strands of evidence, such as the analysis of ice cores to "recreate" atmospheric composition over many thousands of years and relating that to other evidence of surface temperature. The research is undoubtedly challenging given all that other factors that influence temperature, such as solar activity, but the evidence for a human influence on climate change is strong. Assessing the impacts of such a rise in global temperature is very hard to determine, especially when it comes to local scales, and they are likely to vary across the planet with some places experiencing lower temperatures than usual while others face higher temperatures.

The answer to human-mediated climate change is clearly to reduce the level of GHGs in the atmosphere, but there are costs associated with such a change, and this brings us back to the more general points made above about perceptions. Reducing GHG emissions may be seen by some as a price they do not want to pay if they think it results in a lowering of their living standards, and this sentiment can be (mis)used by politicians seeking election. These are deep waters, but given the importance and indeed the urgency of addressing global warming, it is perhaps understandable that there exists something called the Climate Perceptions Index that attempts to capture the public's awareness of climate change, their perceptions of risks associated with climate change and, critically, their commitment to action (van Valkengoed et al., 2021; Htitich et al., 2024). The index was created by a USA-based not-for-profit organisation called the Social Progress Imperative (SPI), and they use data collected by Meta, the technology company behind the social media platform called Facebook, via its annual Climate Change Opinion Survey to populate the indicator. The following quote is from the Climate Perceptions Index report of 2022:

> The Climate Perceptions Index is a comprehensive measure that assesses public perceptions of climate change. Based on data from over 100,000 active Facebook users across 107 countries, this index explores three key dimensions: awareness of climate change, risk perception, and commitment to action. The Climate Perceptions Index offers in-depth knowledge of the societal implications of climate change. It aims to provide insights for political leaders on where to focus to improve public support for climate action in their country. The index covers 107 countries and 3 additional regional groupings of countries.
>
> *(Htitich et al., 2022; page 1).*

The use of a social media platform (Facebook) as the vehicle for obtaining the data for the index is interesting, and the role of social media in this space will be returned to later in Chapter 5.

The full details behind the calculation of this indicator can be found in Htitich et al. (2024) and do not need to be repeated here although a summary is provided as Table 2.1. A more general discussion of the methodology behind perception-based indicators, and the issues that can arise from that, is provided within Chapter 3. The Climate Perceptions Index has three components (awareness, risk perceptions and commitment to action) with each having an equal weighting. The arithmetic mean of respondent responses for each of the countries in the survey is used to generate the value for the Climate Perceptions Index and to allow for easier interpretation;

the scores are "standardised" to fit a range from 0 to 100 with higher values representing a "better" perception of climate change where "better" in this context refers to answers that suggest a higher awareness, stronger perception of the risk involved and a stronger commitment to action. Results presented in the latest version of the Climate Perceptions Index (Htitich et al., 2022) suggest that there are complex patterns of country performance within the three components, but the country at the top of the ranking is Portugal, although the value of the indicator was just 68.5, and the country at the bottom is Yemen with a value of 30.6, representing a somewhat compact range in climate perception from top to bottom of the rankings. It is worth reiterating that all the key decisions—the questions to ask, their weighting, their aggregation and adjustments—are made by the people who create the indicator. Change any of these decisions and the values of the indicator may also change, and this could have significant ramifications for the country rankings.

The engagement of the public, or part of it, is an intriguing approach to populating these indicators, and the point will be returned to in later chapters. It has

TABLE 2.1 Questions that form the basis of the Climate Perceptions Index

Component	*Questions (summarised)*	*Answers (score)*
Awareness	*How much do you know about global warming or climate change?*	I know a moderate amount (0.5) I know a lot (1) All other answers are given a score of 0
	Do you think climate change is happening?	Yes (1) No (0)
	Assuming that climate change is happening, do you think it is caused mostly by human activities?	Yes (1) No (0)
	About how often do you hear about climate change in your daily life?	At least once a week (1) All other answers are given a score of 0
	Maximum contribution to index	**4**
Risk perceptions	*How worried are you about climate change?*	Somewhat worried (0.5) Very worried (1) All other answers are given a score of 0
	How much do you think climate change will harm you personally?	A moderate amount (I0.5) A great deal (1) All other answers are given a score of 0
	How much do you think climate change will harm future generations of people?	A moderate amount (0.5) A great deal (1) All other answers are given a score of 0

(*Continued*)

TABLE 2.1 (Continued)

Component	*Questions (summarised)*	*Answers (score)*
	Do you think that climate change is a threat to the people in the country where you live over the next 20 years?	Somewhat serious (0.5) Very serious (1) All other answers are given a score of 0
	Maximum contribution to index	**4**
Commitment to action	*How important is the issue of climate change to you personally?*	Somewhat important (0.5) Very important (0.75) Extremely important (1) All other answers are given a score of 0
	What level of priority do you think climate change should be given by the government of the country where you live?	High (0.5) Very high (1) All other answers are given a score of 0
	Do you think the country where you live should reduce the pollution that causes climate change?	Regardless of what other countries do (1) All other answers are given a score of 0
	In the country where you live whom do you think is most responsible for reducing air pollution that causes climate change?	The government (1) Businesses (1) Individual people (1) Only one answer is allowed. All other answers are given a score of 0
	Maximum contribution to index	**4**

some loose resonance with the notion of "citizen science" that is often said to have emerged during another environment crisis back in the 1960s linked to "acid rain". This was caused by the emission of pollutants such as sulphur dioxide and nitrogen oxides from the burning of coal and other fossil fuels, which dissolved in water in the atmosphere to make it acidic, hence the term "acid rain" that caused damage to buildings and ecosystems such as forests and aquatic systems (Noël et al., 2021; Aswin Giri and Shiva Nagendra, 2024). One of the most captivating aspects of acid rain was that the impacts often occurred many miles from the places that were releasing the pollution (Likens et al., 1979; Park, 1987; Galloway et al., 1987). National laws and international agreements were introduced to limit the burning of coal, in particular all of which had immediate beneficial results (Grennfelt et al., 2020). But despite the scientific evidence, the introduction of these laws and agreements often did take time. In the USA, for example

> From its discovery in 1963 to passage of the Clean Air Act in 1990, legislative action on acid rain took 27 years.
>
> *(Ogden, 2019)*

Citizens became involved in the monitoring of pollution and its pact on ecosystems, and the term can be traced to an article published in 1989 (Kerson, 1989; Riley and Mason-Wilkes, 2024). Citizen science can take many forms including, for example, the collection and testing of samples as well as observations (Riley and Mason-Wilkes, 2024). It is thus a type of public participation, and numerous authors have made the point that when it comes to environmental management, it is important to engage with people rather than rely solely on input from experts, and such engagement needs to go beyond just helping with the collection of samples or making observations. As Bickerstaff (2004; page 836) has noted:

> many within the social science risk research community have argued for more participatory decision processes to allow a wider range of perspectives to come into play where necessary—not only on moral or ethical grounds but also as a means of widening the expert peer community to check for knowledge gaps and expert biases, as well as to deal with questions of the interaction between preference, ethics and values.

The importance of public participation for the identification and use of SIs will be returned to in Chapter 4.

2.3 What is sustainability, and why do we need it?

The climate change example in the previous section is an illustration of unsustainable practices. The continued burning and generation of GHGs result in climate change, which in turn impacts on all of us, even though the perceptions of some captured via indicators such as the Climate Perceptions Index may suggest otherwise. We can, of course, take no notice of the damage or perhaps not care and continue to generate GHGs until eventually tipping points are exceeded, climate change causes untold levels of damage and our living standards, maybe our very survival as a species, suffer as a result. The untold damage that unfettered climate change will do to us is the very epitome of practices that are unsustainable. We do not need to define unsustainable in this context; we will certainly know it when we experience it, and indeed, there are many who are already living with the outcomes of global warming! But rather than wait for these outcomes of unsustainable practices to impact upon us, how can we know whether practices are sustainable? Moving away from GHG emissions would certainly be a move in the right direction, but is there a level of emissions that avoids global warming? Is it really a case of dropping emissions to zero or is there a level that can be sustainable? In other words, what does sustainability look like?

There are numerous definitions of sustainability, and indeed, some of these make a distinction between sustainability and sustainable development, with the latter seen by some as a more "human centric" conceptualisation of the former. Others regard sustainability as a process while sustainable development is the endpoint.

I am not entirely convinced by these arguments, not least because both are seeking to describe something that is inevitably human-centric; the actions of people and their motivations are at the heart of sustainability and sustainable development. Nonetheless, I will resist the temptation to travel that road and will instead take the commonly held view of seeing the two terms as synonymous. There does seem to be a broad consensus on the following:

> Humanity has the ability to make development sustainable to ensure that it meets the needs of the present without compromising the ability of future generations to meet their own needs.
>
> *(WCED, 1987; page 13)*

This definition of sustainable development is an old one, first published in April 1987 by the World Commission on Environment and Development (WCED), which had been set up in 1983. The definition appears in a WCED report with the evocative and catchy title "Our common future" (WCED, 1987) and is often referred to as the "Brundtland Commission" named after the Chairperson of the WCED—Gro Harlem Brundtland—who at the time was the Prime Minister of Norway and active in various environmental initiatives. While the definition is old, it is perhaps the closest we have to a widely accepted vision of sustainability and is often repeated, and sometimes even slightly modified. Nonetheless, other definitions of sustainability are available, and some can be found in Kuhlman and Farrington (2010), Barbosa et al. (2014), Ramsey (2015) and Ruggerio (2021). Many of these alternatives focus on the need for systems thinking or perhaps on maintaining and improving our quality of life and well-being without depleting the Earth's natural resources and environment. But while definitions of sustainability abound, there are also real dangers of it becoming rooted in abstractions, and it can even become a target for satire and lampooning, as wonderfully illustrated by a graph in XKCD—a webcomic of romance, sarcasm, math and language (XKCD, 2025). The cartoon in the comic comprises a graph showing the extent of use of the word "sustainable" in published texts (vertical axis), calculated as a percentage of all words published in US English each year, since the 1950s (time is used for the horizontal axis). The graph presents a gradual increase in the use of the word "sustainable"; over time and based on a projection of the trend up to 2018, it suggests that by 2026, the word "sustainable" will appear at least once on every page of text that is published in the English language, and by 2061, it will appear once in every sentence published! Protecting even further to 2109, "sustainable" will be the only word that is ever published anywhere and by anyone in English! Such satirical cartoons may be amusing, but the sentiment seemingly being expressed here is that sustainability is growing so rapidly in popularity that it eventually becomes vacuous and meaningless as everyone repeats it for the sake of it. Indeed, this sentiment may provide cover for those that wish to continue to exploit the natural environment, its resources and people; if sustainability is meaningless, then why try to achieve it?

While definitions of sustainability may be useful for framing a broad ambition—a sense of where we want to get to expressed as a positive rather than just saying what we want to avoid—we still need to provide clear guidance for steering actions, and these may need to be quite context (place and time)-specific. Hence, there has been a strong move towards developing frameworks—sets of actions, targets and indicators—to help root the concept of sustainability in a more practical domain, and one of these, the SDGs, will be set out later in the chapter. This "rooting" of sustainability in terms of more practical guidance and assessment has the advantage of making it more engageable for all of us even at very local scales. Indeed, many have called for less emphasis on trying to define sustainability in a sentence and more on setting out practical guidelines and frameworks for its implementation, including the use of indicators.

2.4 Making sustainability happen: Need for targets and indicators

All indicators inevitably involve a trade-off between what could be called their technical and theoretical purity and usability. This is arguably an ineviTable feature of indicators as they are intended to provide an idea of what something is like.

(Collins Dictionary)

for helping to inform action; they are by design meant to be practical tools. But an "idea" of what something is like does sound rather vague, but it may well be the case that having some indicators to guide action may be better than having none even if there are also dangers of using the wrong indicators or interpreting them in ways which are inappropriate. To help address the potential downsides, mnemonic-based frameworks have been developed that set out their requirements and two of them—SMART and SPICED—are covered below. First, an ideal indicator is often said to be one that is SMART; where SMART is an acronym for the following characteristics (Doran, 1981):

- Specific: The indicator should be focussed and accurately describe what needs to be measured. However, what "accurate" means in practice is, of course, open to some debate.
- Measurable: The indicator needs to be given values (or "populated") in such a way that allows for consistency. The logic here is straightforward. If the methodology is not consistent, then it will unclear whether changes over time and space are real or simply down to changes in the way data are collected and manipulated.
- Achievable: The data needed to "populate" (i.e., give values to) the indicator should be straightforward and cost-effective to collect and present. In some cases, the data may need to be combined or transformed in certain ways to make them amenable to presentation and interpretation.

- **R**elevant: The indicator needs to be relevant to the "something" it is trying to capture.
- **T**ime-bound: Indicators are specific to the time when they are collected, and thus, it is necessary to make sure that the values are timely to make them usable. There may be little point in using data collected 10 years ago to populate an indicator that is meant to inform action now. The identification of trends may also require an indicator to be populated with data collected at different times.

The SMART attributes of indicators noted above are a useful starting point, but there are elaborations on this approach. One version similar to SMART is called CREAM, where the mnemonic covers indicators that must be Clear, Relevant, Economic, Adequate and Monitorable (Schiavo-Campo, 1999). Another mnemonic that extends rather than repeats the attributes in SMART is provided by SPICED, where the letters cover the following attributes (Roche, 1999):

- **S**ubjective: Indicators are inherently subjective in the sense that they are created and chosen by people. Any context could perhaps best be captured by a large suite, perhaps hundreds or thousands, of indicators, although there are practical concerns such as logistics or indeed a need to limit information to make it manageable that may result in only a relative few being selected.
- **P**articipatory: Ideally, indicators should be developed by those with good knowledge of the context (the "something") which those indicators are meant to capture. But this may go beyond what are often traditionally thought of as technical "experts" and may include those who live and work in the context or perhaps those who are meant to consume and act on the information captured by the indicators. Indeed, even with just these few examples, the list of potential "stakeholders" could potentially be a large one. There are participatory approaches designed to embrace and condense this diversity, and these will be covered in Chapter 4.
- **I**nterpreted and communicable: This is not covered directly in the SMART set of characteristics, but to have their desired impact, indicators need to be readily understood and capable of being communicated to those who are intended to consume and use them. Having hundreds of indicators may be desirable to cover a specific context, such as sustainable development, but how can those meant to use them work with such a large number of indicators? This brings us to the next characteristic:
- **C**ross-checked and compared: It may well be that a single indicator is not enough to capture the complexity of a particular context, and thus, a number (suite) of indicators may be required perhaps set out as a framework as with the SDG framework mentioned above. This allows for various "windows" on the same context and for them to be cross-checked (triangulated) and compared. It may well be that some indicators are pointing in the same direction, but it is also

possible that the indicators may provide contradictory messages. If the indicators are providing a different picture, then why should this be?

- Empowering: Indicators are tools used to help inform management, but they can also be empowering in the sense that they can act as catalysts to help spur critical reflection, and the information can bring about learning and arm people with new knowledge. These in turn can provide people with a sense of empowerment and encourage their engagement in addressing issues of importance to them.
- Diverse and disaggregated: As noted above, it may be useful to have a set of indicators designed to provide "windows" on the same context, and it may also be desirable to have the same indicator as a "window" on different groups.

SMART and SPICED taken together cover many of the desired attributes of indicators, and these can help guide the selection of a framework. However, it still needs to be remembered that while such attributes can help, there are still many assumptions in all of this, and as noted with the very first attribute of SPICED, there is an inevitable degree of subjectivity whether we like it or not. Much also depends on the quality of the data used to "populate" the indicators. Even a well-conceived indicator may be an undependable stick to lean on if the quality of the data used to populate it is poor.

2.5 SI frameworks

While indicators have a clear appeal, there are nonetheless important issues that need to be considered. Much depends on the choice of indicators and how they are populated and presented, and these choices may also be influenced by practical concerns such as available resources for collecting the required data and communicating the indicators to others. Similarly, indicators may also interact. For example, indicators designed to assess social issues such as healthcare can have a correlation with economic indicators such as average income, and in turn, these can correlate with environmental issues such as air and water pollution. More disadvantaged areas tend to have lower average incomes and greater pollution, and these may result in poorer outcomes in terms of health and perhaps even life expectancy. Given that a group of indicators designed to assess these issues may tend to mirror each other, should we not just focus on average income and forget the others? But there are dangers here as there are likely to be other factors that influence health outcomes besides income, and hence, a strong case can be made to present a group of indicators even if they may have some relationship rather than try and reduce them too much. Indeed, we could simply list all the indicators in categories and not concern ourselves all that much with the nature of relationships between them; in effect, we could just acknowledge that these are present but see the indicators as spanning a range of issues deemed to be important. This latter approach is the one taken by the United Nations with its SDGs.

The SDGs followed on from the Millennium Development Goals (MDGs) set by 189 United Nations member states in September 2000 and agreed to be achieved by 2015. Each of the MDGs had a set of targets and indicators, and while the success of this programme has often been claimed to be mixed (Fehling et al., 2013), it was deemed to be good enough to expand into the much wider SDGs from 2016 to 2030 (Sachs, 2012). While the MDGs were focussed more on the developing world, with targets and indicators geared more towards addressing poverty, its causes and outcomes, the SDGs were intended to be applied by all countries, rich and poor, and incorporate all facets of sustainability (Hak et al., 2016). On 1 January 2016, a set of 17 SDGs came into effect having been adopted by 193 countries in September 2015. This was by no means the first attempt by the UN or indeed others to have global agreements linked to sustainability, but it was the first to be so wide-ranging and to have a structured set of goals, targets and indicators that the countries agreed to achieve.

A list of the 17 SDGs and the distribution of the targets and indicators across them is presented as Table 2.2. In this table, the 17 SDGs have been ordered in terms of the number of indicators they contain, with the highest number at the top of the table and the lowest at the bottom. As would be expected given the breadth of coverage of sustainability, the goals span a highly diverse range of topics from health and well-being (SDG 3) to food security (SDG 2) and energy supply (SDG 7). It may seem somewhat bewildering, but sustainability is by its nature diverse and complex, and while the SDGs are entered in the table as discrete rows, there will inevitably be interactions between them along the lines outlined above. The goals also seem very challenging, and so they have proven to be in practice, but they all matter. Each goal has a set of associated targets, varying from 5 (SDG 7) to 19 (SDG 17), and at the time of writing, the framework comprises a total of 169 targets across the 17 goals with each target typically having one to three associated indicators, which taken all together generates a total of 247 indicators (although 231 of these are "unique" as in some cases, the same indicator is repeated across targets). These are large and perhaps even overwhelming numbers, but again it is important to acknowledge the breadth of ambition here, and when all the goals are considered, the number of targets and indicators does not seem unreasonable given what it is they are trying to address.

The number of targets and indicators for each SDG is uneven. Those having the highest number of targets and indicators are SDG 3 (ensure healthy lives and promote well-being for all at all ages) and SDG 17 (strengthen the means of implementation and revitalise the global partnership for sustainable development). Those with the fewest number of targets and indicators are SDG 13 (take urgent action to combat climate change and its impacts) and SDG 7 (ensure access to affordable, reliable, sustainable and modern energy for all). This balance may be surprising as given the importance of global warming; we would perhaps expect to see more targets and indicators for SDG 13 and the related SDG 7. Similarly, SDG 17 with its emphasis on strengthening partnerships for sustainability seems relatively

TABLE 2.2 The SDGs along with the number of targets and indicators associated with them

SDG number	*SDG title*	*Number of*	
		Targets	*Indicators*
3	*Ensure healthy lives and promote well-being for all at all ages*	13	28
17	*Strengthen the means of implementation and revitalise the Global Partnership for Sustainable Development*	19	24
16	*Promote peaceful and inclusive societies for sustainable development, provide access to justice for all and build effective, accountable and inclusive institutions at all levels*	12	24
8	*Promote sustained, inclusive and sustainable economic growth, full and productive employment and decent work for all*	12	16
15	*Protect, restore and promote sustainable use of terrestrial ecosystems, sustainably manage forests, combat desertification, and halt and reverse land degradation and halt biodiversity loss*	12	14
10	*Reduce inequality within and among countries*	10	14
11	*Make cities and human settlements inclusive, safe, resilient and sustainable*	10	14
5	*Achieve gender equality and empower all women and girls*	9	14
2	*End hunger, achieve food security and improved nutrition and promote sustainable agriculture*	8	14
12	*Ensure sustainable consumption and production patterns*	11	13
1	*End poverty in all its forms everywhere*	7	13
4	*Ensure inclusive and equitable quality education and promote lifelong learning opportunities for all*	10	12
9	*Build resilient infrastructure, promote inclusive and sustainable industrialisation and foster innovation*	8	12
6	*Ensure availability and sustainable management of water and sanitation for all*	8	11
14	*Conserve and sustainably use the oceans, seas and marine resources for sustainable development*	10	10
13	*Take urgent action to combat climate change and its impacts*	5	8
7	*Ensure access to affordable, reliable, sustainable and modern energy for all*	5	6
	Totals	169	247

well-served with 19 targets and 24 indicators. But it should be noted that the number of targets and indicators does not reflect a sense of importance or indeed prioritisation. All the SDGs are regarded as being of equal importance, and the decisions made regarding the targets for each SDG were based on collaboration between experts across the member states of the UN. The selection of indicators was made by a group of experts (the Inter-Agency and Expert Group on SDG Indicators; IAEG-SDGs) using criteria such as availability of data of the required quality, the need for comparability across countries and of course the relevance of the indicators to the targets. The IAEG-SDG group reviews the list of indicators and their values on a regular basis. Hence, the numbers of targets and indicators shown in Table 2.2 are the outcome of all this consultation and collaboration, and no doubt, many assumptions were made, and compromises agreed as the choices were whittled down. It does, of course, mean that those who are not part of the IAEG-SDGs are free to contest all of this and question the assumptions and compromises, and indeed, there are many published critiques of the whole SDG framework including the overly neat compartmentalisation of "topics" into the rows shown in Table 2.2 (Winkler and Williams, 2018; Sneyd, 2024; Orosz, 2025).

It does need to be mentioned that there are alternative frameworks to that of the SDG target and indicator listings, and these include efforts to think more conceptually about the choice of indicators and how they may relate to each other. For example, there are often cause and effect relationships within sustainability, and we may wish to get a handle on both these rather than just consider the effects. Hence, we can think about sustainability in terms of pressures placed on the system and how that effects the "state" of a system, and, if so, then by how much? A classic example here would be the release of a pollutant such as GHGs or those that create acid rain into the environment caused by industrial or agricultural process. We could measure the rate of release of the pollutant (e.g., tonnes per year) as a "pressure" on the system; the higher the rate of the release, then the greater the "pressure". The concentration of that pollutant or perhaps its byproducts in the environment could be measured to provide a window on its "state" (e.g., concentration of the pollutant in water, air or soil). One would expect to see a link between these two as a higher rate of release should potentially lead to more of it in the environment, but this may not necessarily be the case as some pollutants can break down into more harmless chemicals, as well as products that are more damaging. There are also spatial issues to consider as where the pollutant is released may not necessarily correspond to where it ends up being concentrated in the environment; an effect that is relevant for both the GHGs and those pollutants that cause acid rain. Still, there is logic in having a framework that relates pressure and state indicators. Going further, we can also have indicators of impact that capture how the presence of pollutants could cause damage to plants and animals, as well as to human health and well-being. Impact can also be conceptualised in terms of money; the loss of financial value of crops, livestock and trees for example. Finally, we can consider the use of indicators to help assess responses we may implement to address the release and impact of pollutants.

The Pressure-State-Impact-Response (PSIR) framework for indicators was developed to capture this "cause-effect-action" thinking behind sustainability, and the logic has much to commend it (Morse, 2019). It provides a framework for the targeting, planning, monitoring and evaluation of interventions to help address negative impacts on society and promote positive ones. It is perhaps not surprising that the PSIR framework of SIs has remained popular and indeed is often modified to include "driving forces" in society—the bigger forces at play that generate the pressures. For example, an increased rate of release of a pollutant into the environment could be linked to an increasing demand for a particular product or service. For example, greater demand for energy caused by a "driving force" such as industrialisation or maybe an increase in population can lead to more burning of fossil fuels and release of GHGs. Hence, the inclusion of "driving force" indicators is a way of acknowledging that pressures have a context—they emerge out of wider forces at play in society—and this gives us the full story of a Driving force-Pressure-State-Impact-Response (DPSIR) framework as illustrated in Figure 2.1. Indicators could be created which capture change in the demand for energy as a "driving force", which in turn leads to the burning of fossil fuels, the release of GHGs (pressure), higher concentrations of GHGs in the atmosphere (state), as well as the impacts resulting from climate change (e.g., drought, fires and flooding) and indicators to assess the adoption of interventions such as international agreements or maybe adaptations such as flood defences (response).

The right-hand side of the figure represents a cause–effect relationship between indicators; driving force leads to pressure which in turn lead to state and then to impact. The response indicators in the middle capture what can is being done to address the cause–effect chain to limit the impacts at the end.

The DPSIR framework works well when there are clear linkages between the five categories of indicator, as with the energy and climate change example given above. It is easy to see how a demand for more energy as a driving force creates pressures in terms of GHG emissions, concentrations of GHGs in the atmosphere, impacts of these gases in terms of climate change (e.g., flooding) and finally actions to reduce GHG emissions (e.g., renewable energy or maybe even carbon capture) and adaptations. Where it does not work so well is when indicators from the five categories are not so clearly related (Morse, 2019). Human life expectancy, for example, is influenced by many factors and trying to express this as just one pressure would be challenging to say the least. Hence, while the DPSIR framework has its uses, there are occasions where it generates highly complex webs of influence and feedback and may not be suitable. Indeed, given that sustainability embraces the complex interconnections between environment, economy and society, it is therefore unsurprising that the creators of the SDG framework of indicators did not formally adopt it.

2.6 Realities and perceptions: Targets and indicators linked to reducing inequality

Given the large number of SDG targets and indicators, it is not practicable or desirable to go through all of them here, so only one SDG will be discussed as an

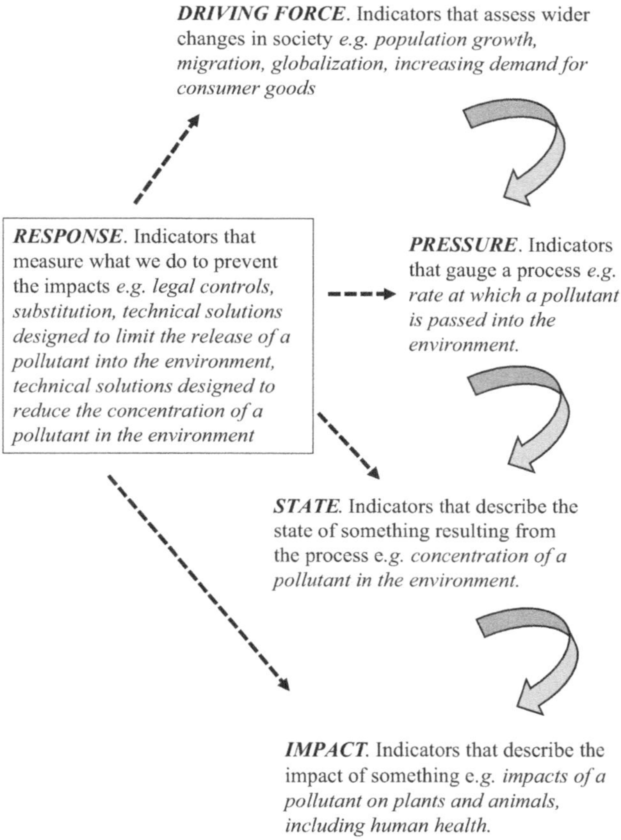

FIGURE 2.1 The DPSIR framework of indicators

illustration. More complete descriptions of the framework can be found on the UN websites plus texts such as Sinha (2020) and Swain and Dobers (2025). The targets and indicators in Table 2.3 are those set out for SDG 10 (reduce inequalities within and among countries). There are ten targets and 14 indicators in total for this SDG, and perhaps, as would be expected from its title, they represent quite an eclectic mix. Some targets (10.1 and 10.2) are based on income, some are about policies and financial regulation of markets (10.4, 10.5), while others are linked to overseas aid (10a, 10b) and migration and mobility (10.7). One of the indicators (10.a.1) covers tariffs applied to imports and is certainly highly topical at the time of writing this book, given President Trump's efforts in the USA to balance what he sees as unfair imbalances in trade by applying tariffs to many countries. Hence, there is certainly a lot to unpack here and while it may be challenging to see any coherent thread that links them the idea is that these seemingly diverse targets and their indicators when taken together to capture the goal of reducing inequality.

TABLE 2.3 The targets and indicators for SDG 10 (reduce inequality within and among countries)

Shaded indicators are those discussed at greater depth in the chapter.

Targets	*Indicators*
10.1 By 2030, progressively achieve and sustain income growth of the bottom 40%of the population at a rate higher than the national average	10.1.1 Growth rates of household expenditure or income per capita among the bottom 40%of the population and the total population
10.2 By 2030, empower and promote the social, economic and political inclusion of all, irrespective of age, sex, disability, race, ethnicity, origin, religion or economic or other status	***10.2.1 Proportion of people living below 50%of median income, by sex, age and persons with disabilities***
10.3 Ensure equal opportunity and reduce inequalities of outcome, including by eliminating discriminatory laws, policies and practices and promoting appropriate legislation, policies and action in this regard	***10.3.1 Proportion of population reporting having personally felt discriminated against or harassed in the previous 12 months on the basis of a ground of discrimination prohibited under international human rights law***
10.4 Adopt policies, especially fiscal, wage and social protection policies and progressively achieve greater equality	***10.4.1 Labour share of GDP*** 10.4.2 Redistributive impact of fiscal policy
10.5 Improve the regulation and monitoring of global financial markets and institutions and strengthen the implementation of such regulations	10.5.1 Financial soundness indicators
10.6 Ensure enhanced representation and voice for developing countries in decision-making in global international economic and financial institutions in order to deliver more effective, credible, accountable and legitimate institutions	10.6.1 Proportion of members and voting rights of developing countries in international organisations
10.7 Facilitate orderly, safe, regular and responsible migration and mobility of people, including through the implementation of planned and well-managed migration policies	10.7.1 Recruitment cost borne by employee as a proportion of monthly income earned in country of destination 10.7.2 Number of countries with migration policies that facilitate orderly, safe, regular and responsible migration and mobility of people 10.7.3 Number of people who died or disappeared in the process of migration towards an international destination 10.7.4 Proportion of the population who are refugees, by country of origin

(*Continued*)

TABLE 2.3 (Continued)

Shaded indicators are those discussed at greater depth in the chapter.

Targets	*Indicators*
10.a Implement the principle of special and differential treatment for developing countries, in particular least-developed countries, in accordance with World Trade Organization agreements	10.a.1 Proportion of tariff lines applied to imports from least-developed countries and developing countries with zero-tariff
10.b Encourage official development assistance and financial flows, including foreign direct investment, to states where the need is greatest, in particular least-developed countries, African countries, small island developing states and landlocked developing countries, in accordance with their national plans and programmes	10.b.1 Total resource flows for development, by recipient and donor countries and type of flow (e.g., official development assistance, foreign direct investment and other flows)
10.c By 2030, reduce to less than 3%the transaction costs of migrant remittances and eliminate remittance corridors with costs higher than 5%	10.c.1 Remittance costs as a proportion of the amount remitted

Also, eight of the ten targets only have one indicator associated with them, while target 10.4 has two indicators and target 10.7 has no less than four indicators.

There is no need to go through all the indicators in Table 2.3, but a couple can be selected for examination, and these are shaded in the table. Indicator 10.2.1 is called "Proportion of people living below 50 per cent of median income, by sex, age and persons with disabilities" and seems straightforward. The indicator comprises the proportion of people having less than half of the median income, and the median is used instead of the mean (average) as distributions of income tend to be positively skewed. The indicator is not based on people's perception of how well off they may feel themselves to be but is instead based on their self-reported income, although the decision to opt for half of the median income as the dividing line is a choice made by those developing the indicator. There is no "law of nature", which says that we must use half-median income as the cutoff point; it could be any percentage we choose. Also, the use of a national rather than international median income means that people below this 50% median income cutoff point in one country may in absolute terms be far worse off than those who are below the 50% cutoff in another country. Nonetheless, this indicator does have an advantage in that many countries will have the necessary income data based on regular surveys and tax returns, and an example is provided for the UK as shown in Figure 2.2. The data in the graphs span the period from 2005 to 2018, and the first of them (see Figure 2.2a) is the percentage of people below the 50% median income, or

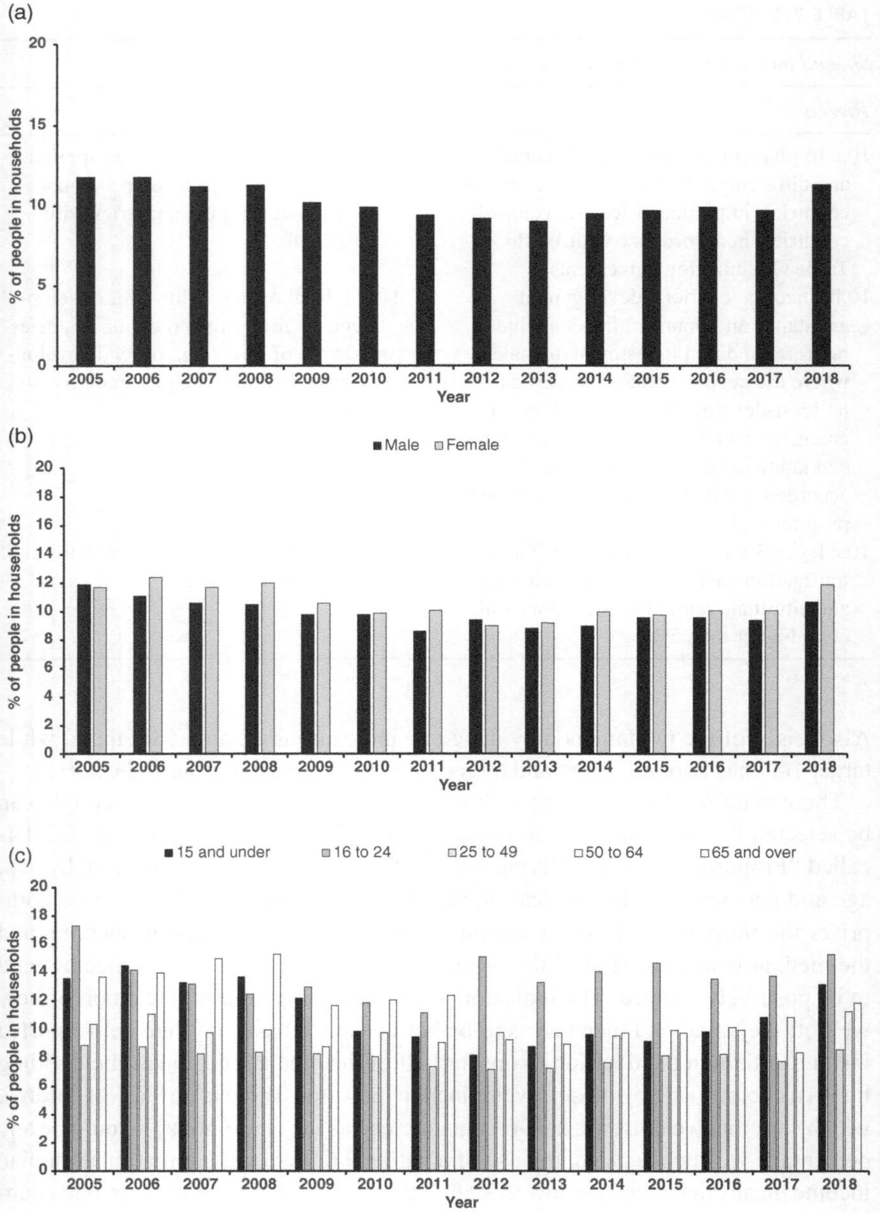

FIGURE 2.2 UK data for SDG indicator 10.2.1—"Proportion of people living below 50 per cent of median income, by sex, age and persons with disabilities"

as the UK government refers to it, the percentage of people in households at risk of poverty. The percentage is relatively consistent at 10–12% but does dip in the period following the financial crash in 2008/2009 before starting to edge up again from 2013. The same data disaggregated in terms of men and women are shown in Figure 2.2b, with the values for women tending to be higher for each of the years. The story for age groups is far more complex (see Figure 2.2c), but in general, the percentages are higher for the youngest and oldest age groups relative to those from 15 to 64 years old, given that this range spans those more likely to be in paid work.

The three graphs show the proportion of household members self-reporting as having less than the half the median income for the UK as a whole and cover the period from 2005 to 2018: (a) The percentage of household members having less than half the median income; (b) the percentage of household members that are male and female having less than half the median income and (c) the percentage of household members for five age groups having less than the median income. There are no comparable data for the UK based on household members self-reported as having disabilities.

Note that there are no data in Figure 2.2 that cover disabilities even though that is specified in the SDG indicator. This reminds us that even for a relatively rich country such as the UK, there can be gaps in coverage. Collecting any data can be expensive, and there can be notable gaps across countries for many of the SDG indicators.

The second indicator of the three is "Labour share of GDP" (10.4.1). The Gross Domestic Product (GDP) is a widely used and reported measure of economic activity. The details of its calculation do not need to be covered here, but in essence it is designed to capture the quantity of money flowing within an economy, and this relates in part to the value of goods and services that are "produced". Hence, GDP can be thought of as a measure of the size of the economy, and given its importance, it is of no great surprise that GDP data are reported regularly by almost all countries across the globe and thus are readily available. The methodologies for calculating GDP can vary across countries or, indeed, over time for a single country, but for the most part, the data are collected via national agencies and forwarded to international ones. GDP often finds its way into political debates and the media avidly, especially at the time of elections, digest and report the values and trends. One of the main ways in which GDP is calculated is via records of expenditure (Morse, 2019). The idea is set out as follows:

$$\text{GDP} = \text{C} + \text{G} + \text{I} + (\text{EX} - \text{IM})$$

where

C is the expenditure of households in the country on goods and services.

G is the expenditure by the pf the country government on goods and services.

I represents investment made by banks (i.e., the use of savings held in banks for generating income).

EX is the sum of exports; goods and services sold outside the country.
IM is the sum of imports; goods and services purchased from outside the country.

It is not necessary here to set out the details behind this calculation, but in broad terms, the greater the spend in the economy by consumers (C) and government (G), then the higher the GDP, albeit with an adjustment that accommodates investment by commercial banks and the balance between exports and imports.

However, while the expenditures approach is a popular one, another way of calculating GDP is via records of income:

GDP = COE + GOS + Taxes Less Subsidies on Production + Taxes Less Subsidies on Sale of Products

where

COE (Compensation of Employees): This comprises payment in the form of wages and salaries to employees.

GOS (Gross Operating Surplus): This mostly comprises profits (and any other income) made by companies providing goods and services.

The other two components are basically measures of income to government via taxation on production (e.g., business rates) and sales (e.g., goods and services sales taxes such as value-added tax (VAT)) but accounting for any subsidies that the government may provide to encourage production and sales.

In theory, these two measures of GDP should match—expenditure in an economy should match income within the economy as there cannot be expenditure without income. However, there may be minor variation resulting from different approaches to measurement. The "Labour share of GDP" assesses the proportion of GDP that goes to workers as wages and salaries and equates to the COE component in the equation as a proportion of the GDP. There are some complications with this such as how to account for those who are self-employed and not on a wage or salary. Hence, reported values of this indicator can vary depending on assumptions made. In the UK, for example, the labour share of GDP has increased from 53.8% in 1996 to 59.5% in 2023. Much does depend, of course, on the availability of reliable data, and in many developing countries, only estimates may be available, and these can fluctuate widely. For example, the estimate for this indicator in Nigeria in 2019 was 46.5%, but in 2006—just 7 years later—the estimate was as high as 90.3% (Source: Federal Reserve Bank of St Louis).

In the context of sustainability, one point worth noting is that the GDP is neutral when it comes to what goods and services are being purchased by households or government or indeed how income is earned. Expenditure by consumers and government may be highly detrimental to the environment and indeed society taken as a whole, but for the calculation of GDP, it is not a consideration. The same may be true when considering income. There may be high salaries being paid to their staff by companies who cause major damage to the environment such as deforestation or large-scale open-cast mining. Both these may cause widespread pollution and negative impacts on quality of life, yet for the calculation of the GDP, this does not matter and indeed may even be a positive! After all, it may result in a need for additional expenditure to repair or cope with the damage. GDP provides a snapshot of monetary flow, which equates to the size of the economy—nothing more and nothing less. There are no value judgements being made here. This issue with GDP and its related suite of like-minded indicators has long been known, and perhaps, one of the best-known and eloquent statements to this effect is provided by Robert F Kennedy in a speech given at the University of Kansas, 18 March 1968. Kennedy was at the time a US senator (New York) running for the Democratic presidential nomination and was assassinated just a few months after making this speech. The indicator he is specifically referring to here is Gross National Product (GNP), a very close relative of the GDP:

> Our Gross National Product, now, is over $800 billion dollars a year, but that Gross National Product—if we judge the United States of America by that—that Gross National Product counts air pollution and cigarette advertising, and ambulances to clear our highways of carnage. It counts special locks for our doors and the jails for the people who break them. It counts the destruction of the redwood and the loss of our natural wonder in chaotic sprawl. It counts napalm and counts nuclear warheads and armored cars for the police to fight the riots in our cities. It counts Whitman's rifle and Speck's knife, and the television programs which glorify violence in order to sell toys to our children. Yet the gross national product does not allow for the health of our children, the quality of their education or the joy of their play. It does not include the beauty of our poetry or the strength of our marriages, the intelligence of our public debate or the integrity of our public officials. It measures neither our wit nor our courage, neither our wisdom nor our learning, neither our compassion nor our devotion to our country, it measures everything in short, except that which makes life worthwhile. And it can tell us everything about America except why we are proud that we are Americans.

This same point has been echoed in numerous speeches, articles and books ever since; it has been made by politicians, economists, academics, teachers, charity workers and indeed businesses. But none of this is the fault of the GDP or the GNP per se; these economic indicators are what they are, and they were only ever intended by their creators to measure the size of an economy in particular ways. GDP was never intended to be a measure of social progress or wellbeing (Coyle, 2014). Some have suggested the adoption, alongside GDP, of other measures that accommodate undesirable and desirable activities for a society, and examples are the Genuine Progress Indicator (GPI) and Index of Sustainable Economic Welfare (ISEW) (Morse, 2019). But these alternatives have never gained anything like the traction of the GDP family of economic indicators, and that begs the question—why? If the GDP is so flawed in capturing sustainability, why is it still so widely used especially as alternatives have long been available?

The final indicator covered here, Indicator 10.3.1, has a somewhat lengthy title—"Proportion of population reporting having personally felt discriminated against or harassed in the previous 12 months on the basis of a ground of discrimination prohibited under international human rights law". It is quite different from the previous two indicators as this one is focussed on what people "feel", and as such, it can be regarded as a perception-based indicator. The focus of the indicator on whether people feel discriminated or harassed is intriguing, especially as the context is "Human Rights Law"—a very broad basis for reflection. It is important to note that the indicator is not based on people's perceptions of discrimination and harassment in wider society but is based on a perception of their own experiences. There is a key difference here, as an individual's perception of the extent of discrimination and harassment in society may be framed by all sorts of factors including discussion with family and friends and the media. But the indicator is specifically based on people's perceptions based on their own direct experience of discrimination and harassment—it is a sort of "self-reporting" in much the same way as Indicator 10.2.1 is based on people's self-reporting of income. Neither is the indicator based on proven incidents; individuals are not being asked to prove that the discrimination and harassment has occurred. Neither are they asked to score the severity of the incidence (did they consider the experience severe or relatively minor?) or where it took place (e.g., work, home, education). The source of data for this indicator are household surveys, typically collected by government agencies, and it is intriguing to note that the results are not required to be subdivided into groups based on gender, age, etc., although this would appear to be optional. This raises the question, of course, as to whether self-reporting of discrimination and harassment is likely to be different from the self-reporting of income? Are people more likely to be truthful when it comes to reporting their income or is vice versa the case?

2.7 Perception indicators in the SDG framework

In the previous section, we explored a couple of indicators SDG 10 (reduce inequalities within and among countries), with one of them being an indicator based on people's perceptions. But how widespread are these perception-based indicators across all 17 of the SDGs? Definitions become critical when trying to make this kind of classification, and here I have divided the SDG indicators into three categories which I have defined as follows. The emphasis in the definitions is very much on how the data to populate the indicators have been sourced, and this information is readily available via the "metadata" indicator sheets provided by the UN:

1. *Reality-based indicators*. These indicators are populated via data collected mostly via government and international agencies. An example would be the

collection of data via tax returns and published financial reports, and the indicator "Labour share of GDP" (10.4.1) is an example of this type.

2. *Self-reporting*. This type of indicator is perhaps the most challenging to define. Here, the indicator is entirely or partially derived from surveys where people are asked to report on their own status, behaviour, achievement, experience and activity. In some cases, expert respondents are asked to report on the progress of their institution in achieving a target. An example here is the "Proportion of people living below 50 per cent of median income, by sex, age and persons with disabilities" (10.2.1), which is based on self-reporting of income via household surveys.
3. *Perception-based indicators*. The data for these indicators are derived primarily from surveys where people are asked to provide their perceptions. An example is provided by the indicator called "Proportion of population reporting having personally felt discriminated against or harassed in the previous 12 months on the basis of a ground of discrimination prohibited under international human rights law" (10.3.1).

Admittedly, there are overlaps between all three of these categories, but especially between categories 2 and 3. Under the "self-reporting" category, there is an element of perception as respondents are being asked to self-report behaviours and positions (e.g., employment status, family size, salary), and this may not necessarily be truthful; there is always the possibility that figures may be under- or over-reported based on people's perception of social norms. For example, people may under-report their consumption of cigarettes, alcohol and drugs and may even over-state their income, all depending on what they feel are the prevailing societal norms. However, in some cases, such as with census returns, it may be a criminal offense to falsify data.

While taking on board the fuzziness that may exist between these three categories, Table 2.4 shows the results of applying them to the list of SDG indicators. The vast majority (85%) of the 247 indicators can be classified as category 1—reality-based indicators. The self-reporting group makes up a further 11% of the indicators, and just 4% of the SDG indicators can be classified as being perception-based. Given the importance of public perception illustrated earlier for environmental issues, it does naturally raise the question as to why perception-based indicators are so poorly represented in the SDG framework? This is a question that will be returned to in the final chapter, but in the meantime, further examples of perception-based indicators, many of which are not in the SDG framework, will be covered in the next chapter.

2.8 Summary

SIs are an important set of tools for the "doing" of sustainability. They allow for the expression of sustainability in practical terms and thus avoid what can be rather vague and nebulous definitions. SIs also allow for the monitoring of progress

TABLE 2.4 Types of SDG indicator; non-perception, perception and self-reporting experiences, behaviour and achievement

Goal number	*Title*	*Type of indicator*			
		Non-perception	*Perception*	*Self-reporting*	*Total*
1	*End poverty in all its forms everywhere*	11	1	1	13
2	*End hunger, achieve food security and improved nutrition and promote sustainable agriculture*	13	1	0	14
3	*Ensure healthy lives and promote well-being for all at all ages*	24	2	2	28
4	*Ensure inclusive and equitable quality education and promote lifelong learning opportunities for all*	12	0	0	12
5	*Achieve gender equality and empower all women and girls*	9	0	5	14
6	*Ensure availability and sustainable management of water and sanitation for all*	10	1	0	11
7	*Ensure access to affordable, reliable, sustainable and modern energy for all*	6	0	0	6
8	*Promote sustained, inclusive and sustainable economic growth, full and productive employment and decent work for all*	12	0	4	16
9	*Build resilient infrastructure, promote inclusive and sustainable industrialisation and foster innovation*	12	0	0	12
10	*Reduce inequality within and among countries*	10	1	3	14
11	*Make cities and human settlements inclusive, safe, resilient and sustainable*	12	0	2	14
12	*Ensure sustainable consumption and production patterns*	11	0	2	13
13	*Take urgent action to combat climate change and its impacts*	7	0	1	8
14	*Conserve and sustainably use the oceans, seas and marine resources for sustainable development*	8	0	2	10

(*Continued*)

TABLE 2.4 (Continued)

Goal number	*Title*	*Type of indicator*			
		Non-perception	*Perception*	*Self-reporting*	*Total*
15	*Protect, restore and promote sustainable use of terrestrial ecosystems, sustainably manage forests, combat desertification and halt and reverse land degradation and halt biodiversity loss*	14	0	0	14
16	*Promote peaceful and inclusive societies for sustainable development, provide access to justice for all and build effective, accounTable and inclusive institutions at all levels*	14	4	6	24
17	*Strengthen the means of implementation and revitalise the global partnership for sustainable development*	24	0	0	24
	Totals	209	10	28	247
	Percentage of total	85	4	11	100

towards the attainment of sustainability; they can be set out in tandem with defined targets and goals as with the SDGs. However, SIs, indeed all indicators, are still human constructs and thus encapsulate what we think are important with sustainability. Even a lengthy framework of indicators, such as that of the SDGs, can still miss much. While the SDG framework is the current dominant force in sustainability, there are other ways of conceptualising SIs rather than as listings of indicators linked to targets and goals, and an example is the DPSIR framework. While the DPSIR framework and indeed other attempts to provide ways of grouping and linking SIs have their advantages, they do also have their disadvantages. Sustainability is, after all, complex and multi-faceted, and there are times when it may be better to take a more pragmatic approach to the use of SIs.

There are many thousands of SIs that could be used, and there are mnemonics such as SMART, CREAM and SPICED to help with the selection of the most appropriate indicators. But even so, it is possible to select indicators that may paint very diverse pictures of sustainability, and at heart, this is a human decision, and much will depend on value-judgement even if the decisions are being made by experts. What may be presented as objective reality may be highly subjective and based on opinion, even if the latter is claimed to be "informed". Change the choices of SIs to monitor and the unsustainable can become sustainable; the just becomes the unjust; the good becomes the bad. As well as the selection of SIs, another major issue is having enough data of the right quality to populate them. Some methods for collecting data to populate perception-based indicators will be covered in the next two chapters, albeit from different angles. With the SDG indicators, for example, there are various types in the mix which vary in terms of how data are sourced, but only a very few of them (4%) are based on capturing people's perceptions although a further 11% are based on self-reporting of behaviour and position. The majority (85%) of the SDG indicators are based on data collected via government returns, monitoring instruments, etc. This disparity and its ramifications will be returned to in Chapter 6.

2.9 Food for thought

1. What do you think of these two approaches (DPSIR and SDG) to conceptualise indicators? Which do you prefer and why?
2. What are the pros and cons of the SDG framework approach?
3. Why do you think UN did not follow the DPSIR framework for their suite of SDG indicators?
4. Why do you think perception-based indicators are so poorly represented in the SDG framework?

References

Aswin Giri J and Shiva Nagendra SM (2024). Air Pollution Perception for Air Quality Management: A Systematic Review Exploring Research Themes and Future Perspectives. *Environmental Research Letters* 19, 053002

Barbosa GS, Drach PR and Corbella OD (2014). A Conceptual Review of the Terms Sustainable Development and Sustainability. *International Journal of Social Sciences* 3(2), 1

Beck U (1992). *Risk Society: Toward a New Modernity*. Sage, Newbury Park, CA

Bennett NJ (2016). Using Perceptions as Evidence to Improve Conservation and Environmental Management. *Conservation Biology* 30(3), 582–592

Bickerstaff K (2004). Risk Perception Research: Socio-Cultural Perspectives on the Public Experience of Air Pollution. *Environment International* 30, 827–840

Collins Online Dictionary. Available at: www.collinsdictionary.com

Coyle D (2014). *GDP. A Brief but Affectionate History*. Princeton University Press, Princeton, NJ

Doran GT (1981). There's a SMART Way to Write Management's Goals and Objectives. *Management Review* 70(11), 35–36

Fehling M, Nelson BD and Venkatapuram S (2013) Limitations of the Millennium Development Goals: A Literature Review, *Global Public Health* 8(10), 1109–1122

Galloway JN, Dianwu Z, Jiling X and Likens GE (1987). Acid Rain: China, United States, and a Remote Area. *Science* 236(4808), 1559–1562

Grennfelt P, Engleryd A, Forsius M, Hov Ø, Rodhe H and Cowling E (2020). Acid Rain and Air Pollution: 50 Years of Progress in Environmental Science and Policy. *Ambio* 49, 849–864

Hak T, Janoušková S and Moldan B (2016). Sustainable Development Goals: A Need for Relevant Indicators. *Ecological Indicators* 60, 565–573

Htitich M, Harmáček J, Lisney J and Krylova P (2024). Multidimensional Cross-Country Measurement of Climate Change Perceptions. Measurement: Interdisciplinary Research and Perspectives 23(4), 343–358.

Htitich M, Krylova P, Harmacek J and Lisney J (2022). *Climate Perceptions Index. Social Progress Imperative*. Available at: www.socialprogress.org
Kerson R (1989) Lab for the Environment. *MIT Technology Review* 92(1), 11–12
Kuhlman T and Farrington J (2010). What Is Sustainability? *Sustainability* 2, 3436–3448
Likens GE, Wright RF, Galloway JN and Butler TJ (1979). Acid Rain. *Scientific American* 241(4), 43–51
Morse S (2019). *The Rise and Rise of Indicators: Their History and Geography*. Routledge, London
MR 89–34. Miscellaneous Recordings, John F. Kennedy Presidential Library
Noël C, Vanroelen C and Gadeyne S (2021). Qualitative Research About Public Health Risk Perceptions on Ambient Air Pollution. A Review Study. *SSM—Population Health* 15, 100879
Ogden LE (2019). *The Bittersweet Story of How We Stopped Acid Rain*. Available at: https://www.bbc.co.uk/future/article/20190823-can-lessons-from-acid-rain-help-stop-climate-change
Orosz A (2025). *Inside the Sustainable Development Goals. From Fighting Poverty to Saving the World*. Routledge, Abingdon
Park CC (1987). *Acid Rain. Rhetoric and Reality*. Routledge, London
Ramsey JL (2015). On Not Defining Sustainability. *Journal of Agricultural and Environmental Ethics* 28, 1075–1087
Riley J and Mason-Wilkes W (2024). Dark Citizen Science. *Public Understanding of Science* 33(2), 142–157
Roche C (1999). *Impact Assessment for Development Agencies: Learning to Value Change*. Oxfam, Oxford
Ruggerio CA (2021). Sustainability and Sustainable Development: A Review of Principles and Definitions. *Science of the Total Environment* 786, 147481
Sachs J (2012). From Millennium Development Goals to Sustainable Development Goals. *Lancet* 379, 2206–2211
Schiavo-Campo S (1999). Performance in the Public Sector. *Asian Journal of Political Science* 7(2)
Sinha S (ed) (2020). *Introduction to the Sustainable Development Goals and Objectives*. Society Publishing
Sneyd A (2024). *Hidden Politics in the UN Sustainable Development Goals*. Columbia University Press
Steg L and Sievers I (2000). Cultural Theory and Individual Perceptions of Environmental Risks. *Environment and Behavior* 32(2), 250–269
Swain RB and Dobers P (eds) (2025). *Routledge Handbook of the UN Sustainable Development Goals Research and Policy*. Routledge, Abingdon
University of Groningen and University of California, Davis (2025). *Share of Labour Compensation in GDP at Current National Prices for Nigeria [LABSHPNGA156NRUG]*. FRED, Federal Reserve Bank of St. Louis, April 27, 2025. Available at: https://fred.stlouisfed.org/series/LABSHPNGA156NRUG
van Valkengoed AM, Steg L and Perlaviciute G (2021). Development and Validation of a Climate Change Perceptions Scale. *Journal of Environmental Psychology* 76, 101652
Winkler I and Williams C (eds) (2018). *The Sustainable Development Goals and Human Rights. A Critical Early Review*. Routledge, London and New York
World Commission on Environment and Development (1987). *Our Common Future*. Oxford University Press, Oxford
XKCD (2025). *'Sustainable'. XKCD. A Webcomic of Romance, Sarcasm, Math, and Language*. Available at: https://xkcd.com/1007/

3

INDICATORS BASED ON PERCEPTION

A new frontier to assessing sustainability?

3.1 Introduction

Indicators are tools that allow us to capture complexity and simplify it in a way that facilitates our conceptualisation of sustainability but also the planning and evaluation of interventions. Indicators can help make solidify what we mean by sustainability and what matters to us in its attainment. Also, as highlighted in Chapter 2, indicators can help provide guidance as to what to do or perhaps what not to do in order to make sustainability happen. The SDG framework is the most recent and perhaps most important set of goals, targets and indicators to help make sustainability a reality across the globe, and as shown in Chapter 2, some, albeit not many, of those indicators are based on our perception of the world around us. Indeed, this is not all that surprising given that perception of environmental issues and indeed all the issues that are wrapped up in sustainability have long been acknowledged as being of importance, and the example of the Climate Perceptions Index was outlined briefly in Chapter 2 as an example. Hence, having indicators based on perception in the SDG framework would make much sense, and the surprise is perhaps that there are not more of them.

One feature of indicators based on perception is that they depend on people's views and this can vary significantly. Asking people to apply a framework such as the Okta scale with detailed guidance as to what comprises each of the points on the scale will almost inevitably generate a range of values for a group of people for the same sky; what one may see as 1/8 of the sky covered in cloud may not necessarily agree with what another may see. The challenges of developing and using indicators based on perception would seem to be immense, but there are many of them that span many aspects of our humanity. In some cases, perception-based indicators are applied to activities that are hidden and relatively inaccessible and

DOI: 10.4324/9781003456605-3

that raises additional challenges. Corruption, for example, causes major social and economic damage (World Bank, 1997; Doig and McIvor, 1999; Hisamatsu, 2003), yet it is very hard to measure in any objective sense as by its very nature it is hidden (Bello y Villarino, 2021). Those benefitting from corruption are hardly likely to provide details on the benefits the transaction provides to them and the tricks they employ. Similarly, those paying the price of corruption may also be unwilling to provide details regarding what they paid and for what benefit. Maybe they would feel ashamed, maybe they would fear prosecution for doing something illegal or maybe they would be afraid of retribution from those they pay. Whatever the reasons, how can we hope to know something that those involved, givers and receivers, do not wish to reveal?

There are also occasions when it may be useful to assess perceptions alongside other indicators that, using the terminology introduced in Chapter 2, that are more "reality-based" as they record actual events. An example here would be the indicators for assessing the crime rate as discussion in Chapter 1. We know that perceptions of being a victim of crime may be quite different to the crime rate as measured via incidents of crime reported to the police (Manning et al., 2022). Not all crime will necessarily be reported for a variety of reasons including fear or trust in the efficacy of the police. People are willing to express their view as to the level of crime in their neighbourhood and what they see as their chances of becoming a victim, and such perceptions of the crime rate can be important in terms of influencing behaviour (Manning et al., 2022).

Indicators populated by gauging perception are important and needed, and understandably, there are many of them even if they are poorly represented in the SDG framework. This chapter will build on the examples of SIs covered in the previous chapter and provide more examples of indicators based on perception. What is it that makes a perception-based indicator tick? What are the areas that have seemingly provided the greatest scope, and indeed demand, for these indicators and why should that be? Finally, how are these indicators given values (i.e., how are they populated)? Perception-based indicators are certainly not new, some have been around for decades, so how do we apply values to them? In many cases, the values are numerical and often based on scales such as the Okta example above, but how is this done for feelings such as happiness or a sense of security and what are the benefits and limitations of such an approach?

3.2 Examples of perception-based indicators

Some examples of perception-based indicators are provided as Table 3.1, and the list is by no means exhaustive as there are many others that could have been included. I do not wish to imply in any way that those selected for Table 3.1 are "better" than any others. Instead, my choice is founded more on the attributes of the indicators and how these provide food for thought. First, I have tried to include a range of such indicators that span different fields (economics, business, geography,

TABLE 3.1 Some examples of perception-based indicators

Name of indicator/index and acronym	*Creator(s) and year of introduction (if applicable)*	*Notes*
ASEAN Competition Business Perception Index (ACBPI)	2019 (guidance published in 2018)	Assesses perceptions of the business community concerning competition in the Association of Southeast Asian Nations (ASEAN)
Beauty and Appearance Indicators	Various indicators and criteria	Indicators of human appearance and personality.
Art Appraisal		Frameworks and criteria to help guide the appraisal of art
Corporate Perception Indicator	CNBC and Burson-Marsteller. First version published in 2014	Based on a survey of business executives and the general population
CPI	TI. First published in 1995	Based on surveys of people's perception of corruption
Consumer Confidence Index	Various formats. One version began in the USA in 1967	Assesses the confidence of consumers
Business Confidence Index	Organisation for Economic Cooperation and Development (OECD) version from mid 1970s	Assesses the confidence of businesses
Global Freedom Scores/Internet Freedom Scores/Democracy Index	Freedom House Published "Freedom in the World" reports since 1973	Variety of data sources but ultimately based on expert evaluations
Global Soft Power Index	Brand Finance version first published in 2005	Assesses perceptions of nation brands
Gross National Happiness Index	Created by the government of Bhutan. First published in 2012.	Comprises 33 indicators spanning nine domains
Gross National Well-being Index	Created by the International Institute of Management in 2005	Tracks seven subjective and objective development areas with no religious component
Happy Planet Index (HPI)	Introduced by the New Economics Foundation in 2006	Includes happiness as one of the components
Social Progress Index (SPI)	Created by the Social Progress Imperative, a US based not-for-profit organisation, and first published in 2011.	Includes perceptions of crime and corruption.

(Continued)

TABLE 3.1 (Continued)

Name of indicator/index and acronym	*Creator(s) and year of introduction (if applicable)*	*Notes*
Global Peace Index	Created by the Institute for Economics and Peace (IEP). First published in May 2009	Includes level of perceived criminality in society
Indicators of purchasing decision	Many examples, mostly from the private sector	
Inflation Perception Indicator	First published in 2004 (Brachinger, 2008). European Commission began to publish inflation perception as of 2019	Inflation as perceived by households rather than the official inflation estimate based on changes in price
Better Life Index	Created by the OECD. First published in 2011.	Includes level of happiness (life satisfaction) and the quality of any social support network
Landscape Perception Indicators	Various indicator frameworks	Designed to capture people's visual perception of the landscape
Net Promoter Score (NPS)	Fred Reichheld in conjunction with Bain & Company and Satmetrix (2003)	Customer likelihood of recommending a company, product, or a service to someone else
Perception of Equal Opportunities	Number of indicators developed by the OECD with the support of Amandi	Spans perceptions of inequality and its causes
Perception of Food Consumption Adequacy	Various indicators and agencies. Including World Bank's Living Standards Measurement Surveys (LSMS)	Perceived assessment of individual or household food security situation
Perceptions of Crime/Safety	Many indicators and frameworks	
Perceptions of Local Area indicators	Various	Typically, a suite of indicators spanning perceptions of the area including appearance, crime, employment, housing, belonging
Risk Perception (individual, organisation)	Many indicators and frameworks	
Risk Perception (global)	Various. For example, the Global Risks Report (GRR) produced by World Economic Forum and the World Risk Poll (WRP) run by the Lloyd's Register Foundation	GRR is based on the perceptions of experts WRP assesses risk based on both perception and personal experience

(Continued)

TABLE 3.1 (Continued)

Name of indicator/index and acronym	*Creator(s) and year of introduction (if applicable)*	*Notes*
Safety Perception Index (global)	Developed by the IEP. First released in 2022	Uses data from the WRP, a global survey designed by the Lloyd's Register Foundation and administered by Gallup
Safety Perception Indicator (personal)	Various indicators.	For example, the UK includes this within its Opinions and Lifestyle Survey (OPN)
Social Responsibility Indicators and Corporate Social Responsibility indicators	Various indicators and frameworks	
Uncertainty Perception Indicator (UPI)	UPI developed by DoCMA and first published in 2020	Both based on an analysis of uncertainly as reported in the media
Economic Policy Uncertainty Index (EPU)	EPU developed by Baker, Bloom and Davis and first published in 2016	
Wellbeing Index (personal)	Many examples. One of them is the online self-assessment tool invented by researchers at Mayo Clinic (USA and UK)	Originally developed to address physician burn-out and distress internally at Mayo Clinic
Numbeo's Quality of Life Ranking	Numbeo, based in Serbia, uses an online crowd-sourcing approach to collecting data	Includes perceptions of crime as well as friendliness and courtesy of healthcare staff
Workplace Perception Indicators	Many examples throughout the private and public sectors	
Employee Perception Indicators		
World Justice Project Rule of Law Index	Created by the World Justice Project. First piloted in 2008.	Based on a set of five questionnaires administered to experts and the general public.
Worldwide Governance Indicators (WGI)	Created by the World Bank and first released in 1999	WGI feature six aggregate governance indicators with data derived from over 30 sources including surveys

governance, social, cultural), all of which have some relevance to sustainability although the angles I am exploring here may not explicitly be part of the SDG framework set out in Chapter 2. In some cases, the indicators will be based on a combination of data based on perceptions (often derived from surveys) as well as other data from other sources, and in some cases, different indicators share the same data sources. Another interesting aside with some of the topic areas listed in Table 3.1 is that they are not really indicators per se but instead are frameworks for assessing aspects of life—beauty and art—that do not use the term "indicator", but which nonetheless appear to have much in common. At first glance, this seems counter-intuitive; how can it be possible to assess something so subjective as beauty and art, and what has this got to do with sustainability? There are so many influences at play here that make any kind of seemingly objective assessment of them impossible. But it has nonetheless been tried, and I will argue that the experiences and insights gained have intriguing overlaps with the development of perception-based indicators in other fields.

It is not necessary to go through all the indicators in Table 3.1 in detail but many of those listed do provide intriguing angles that warrant unpacking. I have chosen to compromise by providing an overview of some of the indicators in Table 3.1 that are especially noteworthy for various reasons. I have also opted to go into greater detail for two of them—the CPI and the suite of indicators created to assess happiness. These two have been chosen for very different reasons. The CPI was selected as it is arguably the best known of all the perception-based indicators and has been created based on numerous assumptions, some of which have changed over time. Indeed, given its popularity, the CPI is perhaps the archetype of a perception-based index and for that reason alone it warrants attention. The indicators of happiness have been selected as they raise all sorts of intriguing questions regarding definition and methods. Just how can happiness be defined let alone assessed and does happiness relate to other aspects of wellbeing that we can assume to be important?

3.3 Perception-based indicators in economics and business

The first one of the "economics and business" indicators in Table 3.1 to be covered here is the Corporate Perception Indicator, created by CNBC-Burson Marsteller and first published in 2014 (CNBC, 2014; Tidy, 2014). As the name implies, the indicator was designed to capture the perceptions of the public and executives towards businesses, and the first version of the indicator was based on a survey across 25 "global markets" of 25,012 people along with a smaller sample of 1,816 business executives. The survey was intended by the creators of the indicator to span "Main Street to the Executive Suite", and the questionnaire for the survey employed some 70 or so questions spanning a wide range of topics linked to how businesses are perceived by the public. The latter includes, for example, the perceived influence of corporate lobbyists, which act for businesses, on government, as well as perceptions of the gap income between company executives and their employees. The timing of the release of this indicator in 2014, not that many years after the financial crash in 2009 and the "Banker Bashing" that some argue was one of its features (Tidy, 2014), may not be coincidence and indeed the impacts of the crash may have contributed to at least some of the more negative findings that emerged in 2014. For example:

> The gap between executives' salaries and the wages of their average employees is seen as growing around the world, as workers believe CEOs [Chief Executive Officers] care more about getting wealthier than about income inequality. . . . The perception of a growing income gap between CEOs and their average worker can hurt performance.
>
> *(Tidy, 2014)*

Following on from the Corporate Perception Index and the point made in the quotation above about the perceptions of income inequality amongst survey respondents, perception of equal opportunities is a theme that has also been captured by a suite of indicators developed by the OECD with the support of Amundi; a European-based asset management company (Amundi, 2023). Data collection was via a "Risks that Matter" survey in 2022 run by the OECD, and this included what call an "Opportunities module" which collected people's perceptions in the following areas:

- Relative importance of different aspects of inequality.
- Views of respondents on the factors that help shape inequality of opportunity.
- People's views of the role played by different actors play in reducing inequality and the type of policies and actions that need to be prioritised to address inequality.

Results can be accessed via the OECD Centre on Well-being, Inclusion, Sustainability and Equal Opportunity (WISE) website but here is a short extract from one of the reports:

> A majority of respondents (60% on average) believe that hard work is essential or very important in determining one's social position. However, of this majority, only one-fifth believe that hard work is the only key factor to success, while most believe that other circumstances are equally or even more important to get ahead in life.
>
> *(OECD, 2023)*

Hence, while hard work matters when it comes to getting on in life, people perceive that there are many other influences at play including place of birth, ethnicity, sex and religion.

Another important consideration for businesses is how consumers of their products and/or services frame their decisions when making purchases. There has, perhaps understandably given its commercial relevance, been much research in this field across a variety of sectors, and there are many factors at play including need recognition, a perception of pleasure that would come from the purchase (which many or may not be realised once the purchase has been made) and perceptions of brand). Reviews on the indicators that people often use when purchasing "green" (i.e., environmentally sustainable) products and services can be found in Maniatis (2016), Sharma (2021) and Joshi and Rahman (2015), and two of the key factors here are the degree of environmental concern shown by consumers and perceptions of the "green" attributes of the product or service, with the latter often communicated via labelling. Indeed, labelling and how that can help frame a "green" perception is another major topic, with a host of systems in place many of which use a score expressed visually via a colour code. The European Union (EU) eco-label scheme, for example "cover the main environmental impacts of the product and their technical performance, including health, safety, social and ethical aspects, where appropriate" (https://environment.ec.europa.eu/topics/circular-economy/eu-ecolabel_en). Examples of eco-labels in use across many jurisdictions can be found at the "EcoLabel Index" website (https://www.ecolabelindex.com/), and at the time of writing, the "Ecolabel Index" spans 456 ecolabels in 199 countries and 25 industry sectors.

An additional topic of interest when it comes to consumer's decision-making over purchases is how ratings of products and services, using "likes", affect such decisions (Ursu, 2018). The use of such "ratings" based on how people perceive the products and services they purchase, as well as the quality of the purchasing process and after-sales support, has become prevalent in all sorts of contexts, most notably on the websites of online retailers. But the same idea, often using "league-table" style rankings based on the ratings has found its way into all sort of areas such as the provision of education (e.g., school and university rankings) and health-case (e.g., hospital rankings). The influence of these ratings and rankings has attracted much attention amongst researchers. Indeed, many of the indicators listed in Table 3.1 are published alongside rankings of countries as a means of "naming and shaming" poor performance and to encourage those with power to improve matters. This is a complex subject as while the rankings may be based on indicators, the rankings themselves also provide an "indicator" of performance that acts as a magnifier. For example, two countries may have very similar values for an indicator, but there can only be one country at the top of the table, and the "winner" could attract much positive attention in the media and elsewhere. However, much will depend on the indicators that are used for the ranking, and very different ranks can be achieved by selecting different indicators or maybe even different methodologies for the same indicator. The indicators may not necessarily be based on good quality and up-to-date data. For example, there may be a significant time-lag, and published rankings may relate to the situation as it was some years ago rather than now, but if they generate a seemingly excellent performance for one country as the "winner" and another as the "loser", then the influence may be strong Morse (2023).

Consumer uncertainty can have a major impact on businesses; if people feel more uncertain about the future, then they may be less willing to spend money. How uncertain people may feel about the state of the economy can be, and often are, gleaned from surveys, but Müller and Hornig (2020) set out an UPI based not on surveys but newspaper reporting of the economy. This change in emphasis is intriguing, but it does have much logic to commend it. Newspapers, printed or online, remain an important source of information for many people, and the stories they carry, along with information from other media, can help frame a sense of uncertainty in the minds of consumers. There is plenty of nuance here as people may also turn to media sources that they think reinforce their own views, and media itself may seek to reflect what it thinks are the views of its readership. Even so, the media does have an important influence, and this will be returned to in Chapter 5. With their UPI, Müller and Hornig (2020) argue that economic uncertainty arises predominantly from three factors:

> 1) Market-based uncertainty. Such as sudden shifts in market sentiment, the bursting of bubbles, or the spreading of pessimistic expectations. The sources of this type of uncertainty are located in the market itself but cannot be forecasted due to limitations of economic models and data availability.
>
> *(page 4)*
>
> 2) Economic policy uncertainty. These arise from developments in the realm of politics that may have economic consequences. An example is the outcome of the Brexit referendum in 2016 and the trade frictions it introduced.
> 3) Truly exogenous economic uncertainty. These come from factors that are located outside of both the market and the political system. Examples may be evolving technological standards (e.g. car engine technologies), technological failures (like the Fukushima nuclear reactor disaster of 2011), earthquakes, volcanic eruptions, pandemics (such as Corona), meteor impacts, severe weather events (e.g. major floods, hurricanes, droughts).
>
> *(pages 4 and 5)*

They employed an unsupervised text mining technique called Latent Dirichlet Allocation (LDA; Blei et al., 2003; Müller and Hornig, 2020; Müller et al., 2021) to assess the quantity of uncertainty-related news in stories published in two German newspapers between January 2008 until March 2020; a total of 752,000 articles. LDA allows for the grouping of articles into clusters having related content, and for the UPI, they noted (emphasis added by myself):

> Naturally, news-based indicators can only find what editors at media outlets considered worth reporting. That's why we call the UPI a perception indicator. What we **actually measure is the amount of uncertainty-related news stories that**

> **citizens are confronted with**, shaping their perception of reality and their expectations. By accumulating content from several news media over extended periods of time indicators may extract additional information about prevalent issues.
>
> *(Müller and Hornig, 2020; page 6)*

While the UPI is framed as a measure of economic uncertainty based on the three factors noted above, it uses just one form of media. Nonetheless, it may be possible to apply the same LDA technique to other forms of media including social media, and this is a topic that will be returned to in Chapter 5.

Inflation, the degree to which the prices of goods and services change over time, is often measured using data captured from surveys of actual prices rather than people's perceptions of those prices (O'Neill et al., 2017). Inflation is often positive, meaning that the value of goods and services increase with time, but it can also be negative suggesting that they decline over time. While the calculation of inflation is based on changes in price as experienced by consumers and businesses, there is nonetheless a place for people's perceptions of inflation. The latter is based on what people "feel" is happening in terms of the prices they pay rather than the real (measured) change in prices. These may not necessarily be the same, and people's perceptions of price inflation can arguably have much impact, maybe even more impact, on their spending and saving behaviour than the real figures. It is, after all, their lived experienced that matters to people more than what they may see being reported in official statistics and the media. People's perception of inflation can thus be of much interest not only to businesses (producers, retailers, service providers) but also to central banks and government (Abildgren and Kuchler, 2019, 2021). An inflation perception indicator was first developed in September 2004 (Brachinger, 2008), and the European Commission began to regularly publish values for inflation perception as of 2019. One of the main findings from these indicators is that people's perceptions of inflation, and perhaps, the direction in which it is heading may often be significantly higher than the real figures (Abildgren and Kuchler, 2019, 2021). This apparent biasing towards an over-estimation of inflation has important repercussions, but, as with so many indicators that attempt to assess perception, much can depend on how the question is asked. Also, there is the heterogeneity of the sample to consider. For example, to what extent is such an over-estimation a prevalent view across a range of socio-economic groups? Is it more prevalent, for example, in poorer households relative to wealthier households, even if all buying their goods and services in the same places? Does such a distinction matter? It may seem harsh, but if inflation is perceived to be worse than it is by poorer households, then it may make little difference to businesses. Poorer households may have little room for manoeuvre in terms of their purchasing decisions and perhaps have little scope for making more discretionary purchases such as entertainment and holidays. But for those households having more room for changing their purchasing behaviour, a view that inflation is bad and perhaps getting worse can put them off from making discretionary purchases, and this could have significant consequences for businesses.

3.4 Indicators of beauty and art

Of all the indicators I have included in Table 3.1, the ones that may perhaps be most surprising are those that attempt to capture beauty. At first glance, this seems like an impossible task given the old saying that "beauty is very much in the eye of the beholder". How can it be possible to identify indicators that allow for the assessment of something as fundamentally subjective as this, open to so many influences? Nonetheless, this is a topic that has attracted much research, and there are plenty of websites that espouse lists of characteristics (some say indicators) used to judge beauty and appearance of people and even, even more astonishing, how that can be used to judge personalities and behaviours. Apparently:

> It is well established that humans have always preferred beauty over brains, and standards of beauty set by society can impact an individual's overall personality to an extent that is indescribable. The face is the part of the human body from which we infer the most information about others, such as gender, identity, intentions, emotions, attractiveness, age, or ethnicity.
>
> *(Kumar et al., 2023; page 292)*

> It isn't necessarily fair, but there is a substantial body of research demonstrating that being perceived as beautiful or handsome offers some advantages. Physical attractiveness has important social consequences. . . . There seems to be an attractiveness stereotype. People perceived as beautiful are viewed as more competent, kinder, and in general as having more socially desirable traits than those considered less attractive.
>
> *(Seide; 2024)*

Given all these points, we can perhaps understand why efforts have been made to come up with characteristics, or perception-based indicators for want of a better term, to assess beauty in humans, and one example is provided by Seide (2024) and summarised as Table 3.2. There is only one explicit mention of indicators in Table 3.2, namely "Indicators of Personality" such as kindness, but there are other aspects in the table that do seem to lend themselves to the notion of "indicators" as defined in Chapter 1. For example, under "Appearance of health", there are physical features that we may use to help frame a perception such as smooth texture of skin and redness of cheeks and lips. None of these are called "indicators" by the author but arguably that is what they are even if they are more qualitative in nature. Maybe we could even come up with scores for the texture of skin and the redness of cheeks and lips?

The quotation above from Seide (2024) that appears to link appearance to perceptions of competence, kindness, etc., may seem far-fetched, but this has been

TABLE 3.2 Characteristics (or indicators) that are typically used to assess beauty in people (after Seide, 2024)

Attribute (or indicator)	*Beauty equates to*
Symmetry	the extent to which one half of an object is the same as the other half
Appearance of health	Features that give an indication of health and vitality (homogeneity of colour, smooth texture, redness of cheeks and lips)
Indicators of Personality	Socially valued traits such as kindness, contentedness or cheerfulness.
Secondary Sexual Characteristics	Masculine features such as a square jaw, Adam's apple, facial hair and a prominent brow ridge are associated with dominance and handsomeness. The same is true of things such as fuller lips and higher or fuller cheeks in women.
Partner Status	Married, highly desired, previous partner
Others	People who look like you, people who look like your parents

explored in research. For example, Kumar et al. (2023) explored beauty/handsomeness as perceived by females and males in India. Participants in the research were shown photographs of females and males and asked to score them based on a set of indicators spanning personal traits, professional traits, interpersonal interactions and experiences, occupational index and professional hierarchy. Based solely on the physical appearances of the people in the photographs, the results generated sets of perceived male and female personality traits for these people. Looks like appearance does matter!

The appreciation of art is another field where it is hard to imagine the use of indicators. People have been passing judgement on the quality of pictures, and indeed, all forms of art since people first started to create them, so it is hardly a new activity and Carney (1994; pages 13–14) makes the point that:

> Almost everyone at some time or another engages in art criticism. And there is a plurality of critical games. An examination of the diverse practices of art critics fails to reveal any established structure or method, much less any type-relative criticism tied to some analytical sequence.

But how can there be a universal framework or rule book which we can follow to tell us if any piece of art is "good" or not? This has attracted the attention of scholars and the debate over whether it is possible or indeed desirable to do this has been raging for long. The dilemma is neatly set out by Darda and Cross (2022), with added emphases from myself:

> The creation and appreciation of art finds a place in all cultures, serving different social, religious, economic, and political functions. If engagement with

> art is indeed universal, as has been argued, then it seems plausible that the processes underpinning aesthetic appreciation are also shared across cultures. Indeed, evidence suggests that people from different cultures base their aesthetic appreciation on **a common set of features such as symmetry, contrast, colour, brightness, complexity, and proportion**. Nevertheless, on face value, nothing seems more subjective than the human appreciation of art. People differ in their aesthetic preferences—some may like contemporary art, while others have intense negative feelings toward it. Previous work also demonstrates how individual differences such as art expertise, understanding, and knowledge, as well as personality traits influence aesthetic evaluation.
>
> *(Darda and Cross, 2022; page 2)*

In their paper, Darda and Cross (2022) describe the results of research where respondents were asked to assess works of art and dance from Indian and western sources using a set of criteria and a 5-point "Likert scale" ranging from "1" corresponding to "not at all", "2" corresponding to "slightly", "3" corresponding to "moderately", "4" corresponding to "very" and "5" corresponding to "extremely".

1. Familiarity (how familiar is the painting/dance?).
2. Complexity (how complex is the painting/dance?).
3. Evocativeness (how evocative or emotional is the painting/dance?).
4. Abstractness (how abstract is the painting/dance?).
5. Technical competency (how technically competent is the painting/dance?).
6. Beauty (how beautiful do you find the painting/dance?).
7. Liking (how much do you like the painting/dance?).

For "dance", there were two other questions:

8. Reproducibility (how reproducible is the dance?).
9. Enjoyability (how much did you enjoy the dance?).

Likert scales and their ilk have often been used as a means of translating something inherently qualitative such as perceptions to numbers, and some examples of "scoring" have already been provided such as the Climate Perceptions Index. Likert scales will be covered in greater depth later in the chapter, but in the Darda and Cross (2022) study, they were used to explore differences in the perception of art and dance between Indian and western respondents and "experts" and "non-experts". The authors found many similarities between groups of respondents in terms of their assessments and concluded:

> [T]he evidence reported here highlights that both cultural specifics and anthropological universals of human art creation and appreciation exist, and are aesthetically relevant across the fine and performing arts, as well as Indian and

> Western cultures. In an increasingly divided and fractured world where many seek to fortify boundaries based on culture, race, and country, our results point to the potential of art and art experience as a unique equaliser that can bind people together.
>
> *(Darda and Cross, 2022; page 23)*

Nonetheless, and while acknowledging the sentiment expressed in the quote, the list of questions summarised above may appear to be somewhat limited and arguably superficial. For example, the questions ask about familiarity, complexity, evocativeness, etc., without trying to unpack any of them. Not only may your sense of the familiarity or complexity of a piece of art or a dance be very different to mine but we also have a different sense of what these mean. The same issue would presumably apply to all the criteria. Hence, the list almost cries out for a secondary level of indicators to help the appraiser pin down the criteria, to help inform what needs to be looked for to identify familiarity, complexity and so on. But will these secondary indicators then require a tertiary level and so on? It could all start to get very involved.

As an alternative approach, some have sought to first set out principles behind any appraisal of art. Barrett (1994) provides an example based on "interpretation" and sets out 18 principles intended to help guide any assessment, and these are provided in Table 3.3. There is some resonance between this approach and the SDG framework that sets out broad principles that are made more tangible via the targets and indicators. Indeed, it could be argued that many of Barrett's principles could apply equally to the process of developing and using perception-based indicators to assess any of the issues in Table 3.1. There are principles here that state that interpretations are persuasive arguments, feelings are a guide to interpretation and no single interpretation is exhaustive of the meaning of an artwork. Surely much the same can be said of any attempt to assess happiness, well-being or indeed quality of life using perception-based indicators? The appraisal of art seems so inherently subjective, yet the principles in Table 3.3 cry out for more specific guidance as to how the judgement can be made; in much the same ways as the SDG goals need specific guidance in the way of targets and indicators to help make them achievable. For example, principle 3 in Table 3.3 notes how some interpretations of art are better than others, without really telling us how "better" is to be recognised; what would be the indicators for knowing if one interpretation is better than another? Similarly, principle 8 says that "Interpretations are not so much absolutely right, but more or less reasonable, convincing, enlightening, and informative", but again what are the indicators that could help know if an interpretation is "reasonable, convincing, enlightening and informative"? Also, under principle 9 we have "Interpretations can be judged by coherence, correspondence, and inclusiveness", which leads to the need for having criteria (indicators) for assessing these.

While Barrett (1994) sets out some broad principles designed to help guide interpretation, others have preferred to set out a series of steps that can be taken when

TABLE 3.3 A list of 18 principles for the appraisal of art (after Barrett, 1994)

Number	*Principle*
1	Artworks have "aboutness" and demand interpretation.
2	Interpretations are persuasive arguments.
3	Some interpretations are better than others.
4	Good interpretations of art say more about the artwork than the critic.
5	Feelings are guides to interpretations.
6	**There can be different, competing and contradictory interpretations of the same artwork.**
7	Interpretations are often based on a worldview.
8	**Interpretations are not so much absolutely right, but more or less reasonable, convincing, enlightening and informative.**
9	**Interpretations can be judged by coherence, correspondence and inclusiveness.**
10	An artwork is not necessarily about what the artist wanted it to be about.
11	A critic ought not to be the spokesperson for the artist.
12	Interpretations ought to present the work in its best rather than its weakest light.
13	The objects of interpretation are artworks, not artists.
14	All art is in part about the world in which it emerged.
15	All art is in part about other art.
16	No single interpretation is exhaustive of the meaning of an artwork.
17	The meanings of an artwork may be different from its significance to the viewer. Interpretation is ultimately a communal endeavour, and the community is ultimately self-corrective.
18	Good interpretations invite us to see for ourselves and to continue on our own.

appraising art. For example, Carney (1994) identified seven sequential steps which he believed can be applied to analyse a picture (Table 3.4). Steps 2, 3 and 4 would appear to have potential for the development of indicators that could be associated with them, and one can readily envisage a list of questions akin to those of Darda and Cross (2022) with associated Likert scales. However, other steps would appear to be less amenable to the use of indicators, largely because they are based on context (art-historical, declarations made by the artist), but even step 7 has some potential to be captured by indicators if set out as a scale of "aesthetic value" or "value" when compared to other pieces.

In a similar vein to the steps set out by Carney (1994), Van Meel-Jansen (2006) refers to the "Magical Number Five" in art appreciation (Table 3.5) and goes on to describe how these criteria when applied as an analytical framework can be used to make comparisons between different categories of people in terms of how they appreciate the same piece of art. Some of the "Magical Number Five" criteria look amenable to indicators and scoring, and an example is "representation and realism". Others may be less obvious, but even here, it is not inconceivable that indicators based on Likert scales could be developed. Under "emotional expression",

TABLE 3.4 An analytical framework for art as set out by Carney (1994)

Step	*Name of step*	*Notes*
1	Locate the Style	Note the art-historical context and its characteristic features upon which the content of the piece depends.
2	**Descriptive Features and Structures**	Note the descriptive features and structures in the piece. For example, the colours, shapes, arrangements, textures, brush strokes and thickness of lines.
3	**Primary Aesthetic Features**	Presence of any representational, expressive and exemplified features.
4	**Value Features**	These encompass aspects of both form and content. For example, the relationships of features in the picture.
5	Low-level Interpretation	The meaning or content of the picture in a basic form.
6	High-level Interpretation	Brings together the low-level interpretations along with such things as the artist's oeuvre, declarations made by the artist about the work and the art in an historical context.
7	Critical Judgement	Whether the picture has (or lacks) aesthetic value to a degree and whether the artwork has more (or less) value than another.

one could readily imagine a series of questions which asks whether the art makes the observer sad, happy, lonely, etc., given that this is in a similar vein to indicators of happiness covered later in the chapter. Admittedly, the range of "feelings" that would need to be set out as structured questions could be quite large but may not necessarily be insurmounTable.

The page numbers of quotes are those from Van Meel-Jansen (2006).

But as can be imagined, such formalised and almost "scientific" approaches to art appraisal have had their critics, and Clements (1979), for example, comments:

> It seems that the scientific method has been applied to all manner of non-scientific, affective concerns such as social intercourse, religion, and value acquisition. Thus it is not surprising, though it may seem repugnant to many, that an attempt should be made to apply the scientific method to art criticism, an area many hoped would remain a sanctuary, safe from systematizing classification laws.
>
> *(Clements, 1979; p 67)*

Could not the same concerns be raised about the use of perception-based indicators in many fields, but especially those related to assessments of human well-being and happiness? Can we draw lines here and, if so, then where should they be? What would be amenable to this application of "scientific method" to "*affective concerns" and what should remain "a sanctuary, safe from systematizing classification laws*"? After raising these concerns Clements (1979) goes on to provide some advice as to how the criticism of art should proceed:

> for criticism of a work of art, it is essential to analyze the color, shape, space, line, texture, pattern, light; the way these mutually interconnect (e.g., colorful shapes, space-creating lines); and the way they are bound together by the principles of design: rhythm, progression, repetition, balance, and variation. As the literal, technical, and formal features are being inventoried, a bridge between formal analysis and interpretation should develop.
>
> *(Clements, 1979; p 71).*

Perhaps ironically given the first quote from Clements (1979), there are intriguing "tilts" in this second quote at potential "indicators" (colour, shape, space, line, texture, etc.) that do not seem all that far away from those in Table 3.5. Indeed, even

TABLE 3.5 The Magic Number Five of Van Meel-Jansen (2006) as the basis for an empirical appraisal of art

Magical Number Five	*Explanation provided in the paper*	*Example remarks provided in the paper*
Attraction	"Immediate attraction to, or aversion of, a work of art" (page 110)	"just beautiful" "lovely colors" "I would like to live in that house" "because there are sailing boats in it and I love sailing"
Representation and realism	"Content of art and its realistic rendering for aesthetic appreciation" (page 110)	"because there are a lot of animals" "just like a photo—you see every little hair" "a biblical scene"
Emotional expression	"Ability to catch the feelings expressed in a work of art" (page 111)	"they look very sad" "evokes a feeling of loneliness"
Style and form	"Capacity to disengage oneself from the content of an artwork and to concentrate on a multitude of formal aspects, such as contours, composition, chiaroscuro, and the artist's handwriting" (page 111)	"a diagonal composition" "wonderful contrasts between light and dark" "the painter has used broad brushstrokes"
Interpretation	"The perceiver using their imagination for giving meaning to the work, for playing with possibilities, and appreciating the art work for its openness and the opportunity it provides to invent meaning" (page 111)	"use your fantasy" "this could be a lot of things" "maybe it is a symbol" "the painter wants to demonstrate" "as if . . ."

some of the principles (rhythm, progression, etc.) can be set out in terms of criteria for aiding identification. It feels like there are tensions here; an uncomfortableness that art appraisal can be reduced to frameworks but at the same time an urge to do just that.

But do these indicators of beauty and art appraisal matter for sustainability? After all, none of the SDGs and their targets and indicators mention beauty or art, and there is also not much mention of culture, so why would there be a need for these indicators? The first point that needs to be made is that art does matter in sustainability, and this relevance can be pragmatic, for example, as a vehicle to raise awareness and facilitate communication, as well as helping to inspire. For example:

> Art and creativity play a significant role in society and sustainability. For example, art can help raise awareness about environmental issues and inspire people to take action. Through visual art, music, theater, and other creative forms, we can explore the beauty and diversity of the natural world and the impact of people on our environment. Furthermore, creativity is essential to the development of sustainable technologies and systems. By encouraging innovation and out-of-the-box thinking, we can find new, more efficient ways to use resources and reduce waste.
>
> *(Gutierrez, 2023)*

Others have made a similar point. Based on the results of a workshop which included the voices of both artists and sustainability practitioners, Heras et al. (2021) suggest that there are key areas where arts-based approaches can help inform and facilitate sustainability:

> embracing more-than-cognitive aspects of knowledge, improving communication, grappling with power dynamics, shifting relationships to nature, and facilitating the envisioning of alternative futures.
>
> *(Heras et al., 2021; page 1887)*

But how do perception-based indicators fit into this? First, and at a fundamental level, the applications of these assessment frameworks to beauty and art appraisal help illustrate many of the issues involved in trying to measure the highly subjective, and the first quotation from Clements (1979) given above strikes at that very issue. It is challenging to envisage anything more subjective than appreciations of beauty, wherever it occurs, and the attempts to come up with frameworks—and indicators in all but name—to formalise the process is almost bound to raise concerns. Is it non-scientific yet masquerading as science; is it repugnant? These are key questions for all attempts to assess perception, yet the literature on perception-based indicators makes little if any reference to the topic of art appraisal. Second, there are participatory approaches that do make use of art, and an example is provided by the Soft Systems approach in Chapter 4 and its use of "Rich Pictures" (RPs) as a device for aiding analysis of complexity. An issue that often emerges from this is how these RPs can be contrasted across groups? Such an analysis of RPs may be useful when it comes to identifying patterns in RPs across groups, and art appraisal frameworks could have a role here.

3.5 Perception-based indicators for the rule of law and governance

Perceptions of governance and the rule of law feature prominently in Table 3.1 and perhaps that is unsurprising as these are regarded as critical foundations for economic growth and the strengthening of social cohesion and, by extension, sustainability. The World Justice Project (WJP) has a "Rule of Law" indicator designed to compare the rule of law across countries and jurisdictions. The 2023 version of the "Rule of Law" indicator has more than 500 variables aggregated into 44 "sub-factors" which are further aggregated into eight "factors", one of which is referred to as "informal justice". The variables are based on

> assessments of over 149,000 households and 3,400 legal practitioners and experts in 142 countries and jurisdictions, making it the most accurate portrayal of the factors that contribute to shaping the rule of law in a country or jurisdiction.
>
> *(WJP report, 2023; page 183)*

The data needed to populate the indicator come from two sources: a General Population Poll (GPP) and Qualified Respondents' Questionnaires (QRQs). The GPP questionnaire is applied at the household scale and includes a combination of 127 perception-based questions and 213 experience-based questions, and it also collects socio-demographic information of all respondents. The "experience-based" questions are based on self-reporting, and the same issues mentioned in Chapter 2 for this type apply equally here. The QRQs are designed to complement the GPP survey by interviewing law experts (practitioners, academics) to collect

> information on the efficacy of courts, the strength of regulatory enforcement, and the reliability of accountability mechanisms
>
> *(WJP report, 2023; page 183)*

There are also various WGI produced by the World Bank and intended to help researchers and analysts explore patterns in the perceptions that people have about governance across countries and how this may have changed over time. The WGI

> are based on existing data sources produced by more than 30 think tanks, international organizations, nongovernmental organizations, and private firms

> around the world. The data sources (1) produce original primary data using their own well-defined methodologies, (2) report respondents' perceptions of relevant dimensions of governance, (3) cover multiple countries, and (4) are regularly updated. The data reflect the diverse views of tens of thousands of survey respondents and experts worldwide.
>
> *(World Bank, 2024)*

The WGI framework has six "government dimensions" (Kaufmann and Kraay, 2024):

- Voice and Accountability
- Political Stability and Absence of Violence/Terrorism
- Government Effectiveness
- Regulatory Quality
- Rule of Law
- Control of Corruption

The data used to capture these dimensions come from surveys of firms and households along with expert assessments and span 200 economies over the period 1996–2023. The survey focuses on ascertaining people's perception within the six dimensions, and this is justified by the World Bank as:

> (1) perceptions matter, because households and firms make decisions based on their views and perceptions of the quality of governance;
>
> (2) for some dimensions of governance such as corruption that do not leave a "paper trail", data measuring subjective perceptions of corruption or self-reported experiences with corruption can provide valuable insights;
>
> (3) data capturing respondents' perceptions and views can provide essential information on the gap between de jure rules and their de facto implementation;
>
> (4) unlike objective indicators that capture the existence of specific laws, rules and regulations, data capturing survey and expert respondents' views are not susceptible to "gaming" where policy makers target reforms to narrowly change specific measures simply because they are included in an aggregate indicator that they wish to influence.
>
> *(Kaufmann and Kraay, 2024; page 9)*

The detailed methodology behind the construction of perception-based indicators for these WGI dimensions is set out in Kaufmann and Kraay (2024), and the process involves a statistical approach known as an Unobserved Components Model (UCM; also known as "structural time series models"). The details of UCM do not need to be covered here, and the interested reader is referred to specialised publications such as Harvey and Proietti (2005). In short, UCM is designed to allow time-series data to be analysed to detect explanators such as trends and cycles (e.g., due to seasons) in addition to predictor variables. Trends and cycles are different. A trend is a change in a general direction, for example, a steady increase or decrease in a variable over time, while a cycle is a repeated pattern of increase and decrease. Economic variables, for example, often incorporate cyclical effects (e.g., cycles of boom and bust for economic growth) as well as gradual trends over time. As it is based on a statistical estimation of these explanators from a set of data, it is possible to generate "margins of error", and the curators of the WGI stress that these margins of error

> need to be taken into account when making comparisons across countries and over time
>
> *(World Bank, 2024)*

The World Bank also notes that:

> The six composite WGI measures are useful as a tool for broad cross-country comparisons and for evaluating broad trends over time. However, they are often too blunt a tool to be useful in formulating specific governance reforms in particular country contexts. Such reforms, and evaluation of their progress, need to be informed by much more detailed and country-specific diagnostic data that can identify the relevant constraints on governance in particular country circumstances.
>
> *(World Bank, 2024)*

The inclusion of the rule of law in the WGI list is to be expected, but interestingly, the WGI does not use the WJP Rule of Law Index. It is also not surprising to see the control of corruption as an important topic, and this will be returned to in greater depth in the next section.

One other initiative that needs to be covered here given its long history and experience is the "Freedom House" organisation and its annual "Freedom in the World Report", which rates people's access to political rights and civil liberties. Freedom House is a non-profit organisation established during the Second World War (1941) to promote the application of the principles of freedom and democracy (Giannone, 2010), and given the context of the time, it is understandable that this was seen as being of especial importance. Following the end of the war, the work of Freedom House was founded on the Universal Declaration of Human Rights, adopted by the UN General Assembly in 1948, and the focus from then till now has been upon implementation. The scoring behind the country rankings in the "Freedom in the World Report" is expert-driven rather than being based on surveys. The following quote from the "Freedom House" website for the 2024 version of their report sets out how the experts engage in the process (emphases in italics are from me):

> Freedom in the World is produced each year by a team of in-house and external analysts and expert advisers from the academic, think tank, and human rights communities. The 2024 edition involved 132 analysts, and around 40 advisers.
>
> The analysts, who prepare the draft reports and scores, use a *broad range of sources*, including news articles, academic analyses, reports from nongovernmental organizations, individual professional contacts, and on-the-ground research.
>
> The analysts score countries and territories based on the *conditions and events* within their borders during the coverage period.
>
> The analysts' proposed scores are *discussed and defended* at a series of review meetings, organized by region and attended by Freedom House staff and a panel of expert advisers.
>
> The end product represents the *consensus of the analysts, outside advisers, and Freedom House staff,* who are responsible for any final decisions. Although an element of *subjectivity is unavoidable* in such an enterprise, the ratings process emphasizes methodological consistency, intellectual rigor, and balanced and unbiased judgments.
>
> *(Freedom House, 2024)*

There is clearly a strong degree of trust in experts to make the right decisions. The trust is reinforced by terms such as "consistency", "rigor" and "balanced and unbiased judgment" in the quote above. Indeed, this sense of trust in the expert runs through many of the indicators in Table 3.1, especially those produced by reputable international agencies such as the World Bank and United Nations. These agencies are deemed to be neutral and thus have no especial agenda to promote a favourable position of some countries above others, and their reports can thus be believed. But even so, as Freedom House themselves acknowledge, there is much subjectivity in this "expert-led" process along with potential for bias and a degree of "group think" to emerge amongst the selected experts. This is a point that will be returned to in the next chapter on participatory methods, but one can readily imagine that the use of expert-based rankings to populate indicators of "freedom" across countries would be open to criticism and that has indeed been the case (Bollen, 1986; Bollen and Paxton, 2000; Steiner, 2016). Bollen and Paxton (2000), for example, breakdown the source of potential expert bias into four components as shown in Table 3.6. These biases may be conscious or unconscious, and it is possible for more than one of them to apply, so clearly care needs to be taken. As a result, Bollen and Paxton (2000; page 79) conclude:

> Those researchers who wish to pursue the creation of subjective measures should be aware of the potential for bias in their ratings. This analysis has

TABLE 3.6 Some sources of potential expert bias (after Bollen and Paxton, 2000)

Source of bias	*Explanation*	*Outcome*
Method	The provision of information from countries which experts use to make their decisions may be distorted and selective.	Experts would be making their judgements on information that has been filtered.
Expert specific	Experts may differ in their interpretation of the information available to them or indeed what they regard as being relevant or indeed irrelevant.	One expert may see some information as being more relevant and important compared to another expert.
External trait	An expert may be influenced by other traits of a country outside of the information that has been provided to them.	An expert may feel that a country has more "freedom" because it is more mature (i.e., "developed").
Expert measurement	Experts may differ in terms of how they take the information available to them and use it to create the ratings.	If the process involves a panel discussion (or series of nested discussions) at some stage designed to arrive at consensus, there is potential for some dominant individuals to have a high influence and for the "group think" to emerge as noted above.

> shown that, in democracy ratings, the most likely source of bias comes from the inclusion of extraneous factors (ratings of other constructs could have different sources of bias). It is possible that using a panel of judges for ratings, especially judges with diverse orientations or experiences, could reduce the amount of bias in grouped panel ratings. An alternative line of research is to return to "objective" measures of liberal democracy. . . . A final promising route to improving measurement is to look for ways to combine the information from subjective and objective measures of democracy. All of the above solutions suggest that there are alternatives to using unadjusted subjective measures of liberal democracy.

But this keeps open important questions such as how diverse the "*orientations or experiences*" should be and indeed how many experts should be involved? The 2024 version of the report involved 132 analysts and around 40 advisers to generate the scores and hence rankings, and Freedom House do stress the diversity of this group in terms of their region of origin, but even so a degree of bias, even if unconscious, would seem to be likely. Steiner (2016), for example, has noted how the Freedom House scores tend to favour the US and its allies. Nonetheless, can the rankings of "Freedom House" be so lightly discarded? They do have a degree of subjectivity and potential for bias but if acknowledged, as Freedom House have done, then surely there may be some value here even if care needs to be taken to avoid being too granular.

3.6 The Corruption Perception Index

One crime that has attracted much attention in the multiverse of perception indicators is corruption. Indeed, the CPI is perhaps the archetype perception-based indicator. Corruption has long been known to be an important issue, given how damaging it can be for the economy and society (Goudie and Stasavage, 1998; Doig and McIvor, 1999; Lambsdorff, 1999; Hisamatsu, 2003), and there have been various attempts to capture it using surveys. The CPI was created by a non-government organisation called TI with headquarters in Berlin, Germany, and the idea was to rank countries based on the value of the index and thus generate a sense of "name and shame". Countries towards the top of the rankings are those deemed by the assessment process to be least corrupt while those towards the bottom of the rankings are most corrupt. Corruption is everywhere, and no country is free of it, so the value of the CPI ranking is arguably much more about comparisons between countries than the value of the index per se. As TI puts it on their website:

> What does a number mean to you? Each year we score countries on how corrupt their public sectors are seen to be. Our Corruption Perceptions Index sends a powerful message and governments have been forced to take notice and act. Behind these numbers is the daily reality for people living in these countries. The index cannot capture the individual frustration of this reality, but it does capture the informed views of analysts, businesspeople and experts in countries around the world.
>
> *(Source: www.transparency.org/research/cpi/overview)*

TI defines corruption quite simply as:

> the abuse of entrusted power for private gain.
>
> *(Source: www.transparency.org/en/what-is-corruption)*

Sending a "powerful message" is at the very heart of the exercise and the CPI rankings are regarded by some as:

> the first non-commercial global corruption ranking
>
> *(Baumann, 2020; page 504)*

The CPI has been influential, not least because it is widely reported by the media, and this a point that will be returned to in Chapter 5. Indeed, the impact has been described by Gilman (2018; pages S74–S75) as:

> At the beginning of each year, the whole world seems to be watching when Transparency International (TI) releases the Corruption Perception Index (CPI). The media; nongovernmental organizations (NGOs); opposition politicians; and countries that are geographic competitors seem fascinated by a country's ranking. Is my country 31st, 66th, or 99th? Did the ranking go up or down from the previous year? If so, how many places has it advanced or declined? How does a country compare with others in the region? Local media pick up the story focusing on villains or heroes in the corruption story. International wire services pick up the refrain, with articles about triumphs and disasters in the Middle East, Eastern Europe, or South America. Experts give their opinions on how the countries succeeded or failed based on their ranking.
>
> Globally, those in government hold their breath, especially in specialized anti-corruption bodies or in ministries of justice. Some countries have worked diligently to fight corruption and at times find themselves slipping in the index. Other countries seem to be doing little or nothing and yet they move several positive positions upward. As a result, promotions come to some officials while others lose their jobs. Foreign direct investment increases or disappears. Opposition political parties use the CPI as a cudgel. Aid from multinational and government aid organizations ebbs and flows based on the ranking in the index. In a few cases, significant declines in ranking on the CPI contributed to governments falling.
>
> All of this based on a single number.

The CPI is readily and freely available to anyone with an internet connection and the TI release social media and "press packs" when they publish updated rankings each year (Wilhelm, 2002; Davies, 2004; Baumann, 2020). However, the methodology behind the CPI is complex and has changed significantly since its first introduction in 1992. TI do not collect any data themselves and instead rely on survey data collected by others, and these organisations are not collecting their data at the behest of TI. There are various commercial and non-commercial surveys of corruption undertaken each year by a variety of commercial and non-commercial agencies, and TI tap into these sources to populate the CPI. Indeed, as each of these sources produce their own country ranking based on corruption, the CPI is in essence a "ranking of rankings". The 2023 version of the CPI was based on a total of 13 rankings of corruption, and these are summarised as Table 3.7. The choice of what sources to include in the CPI was based on various considerations, including methodological reliability and institutional reputation, cross-country comparability and multi-year data availability. The full details of these sources and how the data were collected and analysed for the 2023 version of the CPI can be found at

Source	*Assessment type and number of countries/territories*	*Relevant questions in survey and scoring*
African Development Bank Country Policy and Institutional Assessment	Expert evaluations 37 African countries	Separate assessment of: (a) accountability of the executive to oversight institutions and of public employees for their performance (b) access of civil society to information on public affairs (c) state capture by narrow vested interests Scale from 1 (very weak) to 6 (very strong).
Bertelsmann Stiftung Sustainable Governance Indicators	Combination of quantitative data and qualitative expert assessment. 41 EU and OECD countries.	*To what extent are public officeholders prevented from abusing their position for private interests?* Scale from 1 to 10; with 1 representing a low level of prevention and 10 meaning a high level of prevention
Bertelsmann Stiftung Transformation Index	Qualitative expert survey 137 countries and territories	Two relevant questions on corruption: *To what extent are public officeholders who abuse their positions prosecuted or penalised?* *To what extent does the government successfully contain corruption?* Scale of 1 to 10. A score of 1 representing little in the way of prosecution/penalisation/containment and 10 representing a higher level of prosecution/penalisation/containment
Economist Intelligence Unit Country Risk Service	Risk evaluations based on qualitative and quantitative indicators 131 countries and territories	Series of relevant questions on corruption. Scores are given on a scale from 0 (very low incidence of corruption) to 4 (very high incidence of corruption).
Freedom House Nations in Transit	Qualitative expert assessment 29 countries and territories	Series of relevant questions on corruption. Scores are given on a scale from 1 (lowest level of corruption) to 7 (highest level of corruption)
Global Insights Business Conditions and Risk Indicators	Commercial business expert assessment 209 countries and territories	Experts asked to score the following: *The risk is that individuals/companies will face bribery or other corrupt practices to carry out business, from securing major contracts to being allowed to import/export a small product or obtain everyday paperwork. This threatens a company's ability to operate in a country or opens it up to legal or regulatory penalties and reputational damage.* Score from 1.0 (maximum corruption) to 5.0 (minimum corruption); allow half-point intermediate scores (e.g. 3.5)

(Continued)

TABLE 3.7 (Continued)

Source	*Assessment type and number of countries/territories*	*Relevant questions in survey and scoring*
IMD World Competitiveness Yearbook	Executive opinion survey (EOS) 64 countries and territories	Respondents asked the question: *Bribery and corruption: Exist or do not exist?* Score from 0 to 10 scale where 0 is the highest level of perceived corruption and 10 is the lowest.
Political and Economic Risk Consultancy. Asia Risk Guide	EOS 16 Asia Pacific countries/territories, plus the United States	Respondents asked the question: *How do you grade the problem of corruption in the country in which you are working?* Scores from 0 (not a problem) to 10 (a serious problem)
The PRS Group International Country Risk Guide	Risk assessment 140 countries	Reviewers asked to assess actual or potential corruption in the form of excessive patronage, nepotism, job reservations, exchange of favours, secret party funding and suspiciously close ties between politics and business. Scores from 0 (highest potential risk) to 6 (lowest potential risk). Allows half point intermediate scores (e.g. 3.5)
Varieties of Democracy Project	Expert survey 179 countries	Question about corruption as follows: *How pervasive is political corruption?* Scores on a continuous scale from 0 (lowest level of corruption) to 1 (highest level of corruption).
World Bank Country Policy and Institutional Assessment	Expert assessment 74 countries	Experts asked to assess: *Transparency, accountability and corruption in the public sector.* Scores from 1 (low levels of transparency) to 6 (high levels of transparency). Allows for half-point intermediate scores (e.g. 3.5).
World Economic Forum Executive Opinion Survey	Business executive survey 91 countries had valid data for the corruption-related questions	Series of questions on incidence of corruption Scores from 1 to 7, where 1 means very common and 7 means never
World Justice Project Rule of Law Index	Expert survey 142 countries	Total of 53 questions on the extent to which government officials use public office for private gain. Scores are given on a continuous scale from a low of 0 to a high of 1.

Most of the data from these sources were collected in 2023 but some were collected in 2021 and 2022.

TI. Some of the rankings had a narrow geographical focus (Africa, Asia), while others were more global, but even with the latter, the number of countries and territories included did vary. For each of the 13 rankings, the data are mostly based on scoring provided by experts, with a variety of scales and "polarities" (i.e., whether higher scores relate to less or more corruption and vice versa). Also, the phrasing of questions asked or statements made about corruption varied between the sources in Table 3.7.

An example calculation of the CPI for Denmark and Somalia is provided as Box 3.1. The process outlined in Box 3.1 is somewhat tortuous and includes an assessment of the CPI as well as a level of "confidence" in its value. The starting point in all of this is an inherently subjective set of data based on the perceptions of samples of people, albeit many of them being "experts" in corruption in the sense that they have direct experience of it. What follows is a series of statistical calculations, and the use of so many decimal points and statistics has a "feel" of making it all seem very "objective", "precise" and perhaps even "scientific". Nonetheless, it is easy to lose sight of the nature of the data at the very heart of it all, and where the data have come from and how they have been manipulated does matter with all indicators. After all, the CPI is used to rank countries in terms of their level of perceived corruption, and changing any of these assumptions (choice of data sources, how data are manipulated, etc.) can potentially change the ranking of countries and any lessons that can be drawn from that (Morse, 2023). Decisions do indeed have consequences.

BOX 3.1 EXAMPLE CALCULATION OF THE CPI

To create the CPI based on the 13 sources, the results for each country are first converted to a common scale of 0–100, where lower values represent higher perceived levels of corruption and higher values equate to lower levels of perceived corruption. The CPI is calculated as a simple average (rounded up) of all the available rescaled scores for that country, and examples of the calculation based on data for the 2023 CPI are shown in Table 3.8 for the country at the top of the rankings (Denmark; least corrupt country) and the one at the bottom (Somalia; most corrupt country).

Table 3.8 shows the reported CPI score of 90 and 11 for Denmark and Somalia, respectively, and these have been rounded up from the arithmetic means (averages) of all the re-scaled scores available for that country. Please do note that the number of decimal points in Table 3.8 for the re-scaled scores and the various statistics may perhaps seem rather excessive as there are five of them, but these are presented here as provided by TI. It does show a rather odd, and arguably unwarranted, level of implied precision for these scores given the nature of the data upon which they are based, but that is what TI has presented, and I have repeated it here to help make that point.

TABLE 3.8 Example calculation of the 2023 version of the CPI for Denmark and Somalia (TI)

Sources of data	*Re-scaled scores*	
African Development Bank (Country Policy and Institutional Assessment)		3.530485
Bertelsmann Foundation Sustainable Governance Index	96.99364	
Bertelsmann Foundation Transformation Index		8.908407
Economist Intelligence Unit Country Ratings	89.67447	
Freedom House Nations in Transit		
Global Insights Country Risk Ratings provided by the	83.25637	22.45197
International Institute for Management Development in their World Competitiveness Yearbook	97.89119	
Political and Economic Risk Consultancy Asia Risk Guide		
Political Risk Service group, International Country Risk Guide	100	6.292267
Varieties of Democracy Project	78.89152	13.78506
World Bank CPIA		10.11772
World Economic Forum. EOS	82.20667	
World Justice Project Rule of Law Index	87.68097	
Statistics		
Number of sources	8	6
Arithmetic mean	89.57435	10.84765
Standard deviation	7.971621	6.661449
Standard Error (uncorrected for finite population)	2.818394	2.719525
Reported CPI score for 2023	90	11
Calculation of 90% confidence intervals		
Finite Population Correction (FPC)	0.645497	0.763763
Corrected SE (for finite populations)	1.819265	2.077072
Critical z-value at P<0.1	1.64	1.64
Lower CI (90%)	87.0164	7.593603
Upper CI (90%)	92.9836	14.4064

Note: The 4–6 decimal points used for the re-scaled scores, arithmetic means, etc., are those reported for the CPI 2023.

Also shown in Table 3.8 are calculations of the standard deviation for the sample of surveys for which re-scaled scores are available and the Standard Error (SE), which is found as:

$$\textit{Standard Error} = \frac{\textit{Standard deviation}}{\sqrt{(n)}}$$

where n is the number of re-scaled scores for that country. Hence for Denmark, this is:

$$\textit{Standard Error (Denmark)} = \frac{7.971621}{\sqrt{(8)}} = 2.818394$$

But strictly speaking, the SE cannot be found this way as the number of surveys available for the countries is a relatively high proportion of the total population of surveys. Some countries in the 2023 dataset have as few as three rescaled scores (e.g., Maldives and the Solomon Islands), while others have ten (e.g., Poland, Hungary, Slovenia, South Korea). Hence, the percentage of surveys available for each country (i.e., the sample) varies between 23% (3 out of 13) and 77% (10 out of 13). The SE calculation shown above assumes that the samples (n) are based on a relatively small percentage of a total population (N), but that is clearly not the case here as the percentages are relatively high. Statisticians argue that for situations such as this where values of n are close to N, a "Finite Population Correction" (FPC) must be applied to the SE. A commonly used "rule of thumb" is that when sample sizes are 5% or more of the overall sample, then the SE should be corrected using the FPC as:

$$\text{Finite Population Correction (FPC)} = \frac{\sqrt{(N-n)}}{(N-n)}$$

and

$$\text{Corrected SE} = \text{SE X FPC}$$

Note that if n is the same as N, then the numerator of the FPC equation becomes zero and so does the FPC and the corrected SE. This makes sense as the SE is a measure of the variation we would expect to see between sample means; the higher the SE, then the greater the variation expected between sample means. If the sample is the whole population (i.e., n is equal to N), then there is no error involved in estimating the population mean as the sample mean is the same as the population mean, and the SE should be zero. Incidentally, without the use of the FPC, it is still arithmetically possible to calculate the SE when n = N, but the value is meaningless.

In the case of Denmark, the corrected SE can be found as:

$$\text{FPC (Denmark)} = \frac{\sqrt{(13-8)}}{(13-1)} = 0.645497$$

Multiplying the FPC by the calculated SE gives a corrected SE of:

$$\text{Corrected SE (Denmark)} = 0.645497 \text{ X } 2.818394 = 1.819265$$

TI have made the decision that calculation of the SE matters for the CPI as they have been keen from the very first CPI in 1995 to provide the value of the index (i.e., the arithmetic average of the re-scaled scores for each country) along with a "confidence interval" (CI); where the CI is a statistical device designed to provide a degree of confidence in the index. In the 2023 version of the CPI, they achieved this by using the corrected SE and multiplying it by the 90% critical value of the standard normal distribution (they assumed a normal distribution for the re-scaled scores), which equals 1.64. Using the values for Demark as an example:

90% confidence interval = 1.64 X 1.819265 = 2.9835946

Upper 90% CI = 90 + 2.9835946 = 92.9836

Lower 90% CI = 90 - 2.9835946 = 87.0164

Again, these decimal places (four of them) are those reported by TI.

3.7 Indicators for happiness

3.7.1 *Gross national happiness in Bhutan*

Happiness is often defined as the "state of being happy", while "happy" is a "feeling or showing pleasure or contentment" (*Oxford Dictionary*), and one famous and often-cited example of the use of indicators to assess happiness is the Gross National Happiness Index (GNHI) of Bhutan (Monaco, 2016). The notion of "gross national happiness" (GNH) in Bhutan dates from the 1970s, so the history is a long one, and it was inspired by the Buddhist concept of "The Middle Path", which is far older (Brooks, 2013). The drive for GNH is summarised as follows (Ura et al., 2012; page 6)

> The goal of Gross National Happiness—or GNH—was first articulated by the Fourth King, His Majesty Jigme Singye Wangchuck. He built upon the legacy of Bhutan's government since the 1729 legal code by Zhabdrung Rimpoche, which dates from the unification of Bhutan. The legal code stated that "if the government cannot create happiness (dekidk) for its people, there is no purpose for the government to exist" (Ura, 2010). The Constitution of Bhutan (2008, Article 9) directs the State "to promote those conditions that will enable the pursuit of Gross National Happiness".

The GNHI is an attempt to translate this state-level policy emphasis on happiness to an indicator that allows the tracking of progress. Full details of the GNHI methodology can be found in Ura et al. (2012) and Alkire and Foster (2008), so only a brief outline is provided here. At the time of writing, the GNHI is based on a total of 38 sub-indexes, 72 indicators and 151 variables that span the following nine domains considered vital for happiness:

1. Psychological well-being
2. Health
3. Education
4. Time use
5. Cultural diversity and resilience
6. Good governance
7. Community vitality
8. Ecological diversity and resilience
9. Living standards

The domains have an equal weight but the number of indicators and hence variables that comprise each of them are not equal. Some of the indicators are based on "real" data (as defined and explained in Chapter 2), while others are based on perceptions derived via surveys; a mixture that is not unusual for the indicators in Table 3.1. The indicators within each of the nine domains are used to create an aggregated indicator of the domain for each respondent.

One intriguing point of the methodology behind the GNHI is its similarity to widely used and reported measures of poverty such as the Headcount Index (the proportion of the population below a nominal poverty line) and the Poverty Gap Index (the average "depth" of those below the poverty line). The key with such measures of poverty is the establishment of a "poverty line" below which people are deemed, by definition, to be in poverty. Typical values for an "international" poverty line used by development agencies are US$1 or US$2 per day and to some extent is a subjective one, although in fairness, the figures are not plucked from thin air but are rationalised based on known distributions of daily income. Nonetheless, values of these poverty-line based measures of poverty are, of course, heavily reliant on the choice of a poverty line, and where that line is drawn can vary across countries. Using this same "poverty line" idea to assess happiness generates the same basic issue; what should be the appropriate "happiness sufficiency cutoff" (i.e., equivalent to poverty line)? For each of the aggregated domain indicators the curators of the GNHI apply what they refer to as a "sufficiency cutoff"; in essence they define a value of the aggregated indicator which is deemed by them to represent sufficiency for happiness. Every value at or above that line is classified as "sufficient" in happiness in much the same way that anyone above the poverty line is deemed to not be in poverty. In does not matter how far above the "happiness sufficiency" people are as they are still deemed to be "sufficient". Just as with the poverty line-based measures of poverty, what really matters are those below the poverty-line. For the GNHI any value below the "sufficiency cutoff" is "insufficient" ("not fully happy"). For those values which are insufficient, we can add up the number of cases where that happens, and we can also determine the gap between the actual value and sufficiency; we can think of this as a "happiness gap" in much the same way as we can conceive and calculate the Poverty Gap Index.

To provide the reader with a more empirical example of the points above, the process is summarised as Table 3.9 with made-up data for just four respondents as reported by Alkire and Foster (2008). Part (a) of the table sets out the data for the four respondents across the 9 domains (only aggregated values are provided for each domain, not the individual indicators) and part (b) provides the gaps between the data in part (a) and the "sufficiency cutoff" for happiness. In some cases, the gaps will be positive—the values are above the cutoff, and these are shaded—while for others it will be negative—the values are below the cutoff. The GNHI is calculated using both the number of cells below the sufficiency cutoff as well as the depth of those respondents below the sufficiency cutoff. There is likely to be significant variation between respondents.

As with the poverty-line approach, the use of a "cutoff" for assessing happiness certainly provides much scope for subjectivity as much depends on where the "cutoff sufficiency" is set. Even for the same dataset, changing the cutoff could dramatically change the values of the GNHI. At a more fundamental level, the key assumption is that happiness can be captured using the 9 domains and the indicators within them, but these all seem very indirect. The assumption seems to be that responding positively in these domains equates to or perhaps sets the basis for happiness, but it is intriguing that the GNHI does not include any element which asks people how happy they actually feel! Despite this absence of a direct measure of happiness, the curators of the GNHI use its value to classify people into categories as shown in Figure 3.1 Anything above a value of 50 for the GNHI is deemed as "happy" while anything below 50 is "unhappy". For values in the "happy" category, there are subdivisions into deeply happy, extensively happy and narrowly happy, but there is no corresponding subdivision for the unhappy category. The decision, of course, as to what values of the GNHI constitute these categories is also a subjective one. The percentages of the sample for 2022 that fell into the categories are also shown in Figure 3.1 (Oxford Poverty and Human Development Initiative, OPHI). It appears that in 2022 the majority (84% or so) of the population are either narrowly or extensively happy; despite these two categories being defined in such a way that the ranges in GNHI are narrow compared to the two extreme categories. This certainly makes for comforting reading for the government of Bhutan, and perhaps the population at large, but it is important to remember how much subjectivity there is in all of this.

3.7.2 Happiness around the world

Another, more global and arguably more direct (at least when compared to the GNHI) approach to assessing "happiness" is provided in the "World Happiness Report" (WHR) which was first published in 2012, a year which coincided with a United Nations High Level Meeting called "Wellbeing and Happiness: Defining a New

TABLE 3.9 Example calculation of the Gross National Happiness Index (GNHI) for Bhutan (after Alkire and Foster, 2008).
The 9 Happiness Dimensions (columns) are those used for the Gross National Happiness Index of Bhutan.
(a) Raw data and sufficiency cutoffs for the 9 domains of happiness

Happiness domain									
Respondent	Psychological wellbeing	Health	Education	Time use	Cultural diversity and resilience	Good governance	Community vitality	Ecological diversity and resilience	Living standards
A	7	8	7	6	8	5	7	8	6
B	4	9	4	5	3	7	6	5	4
C	9	5	5	8	3	3	8	7	4
D	2	9	8	1	4	4	9	3	4
Sufficiency cutoff	5	9	7	5	6	4	8	7	6

The Figures in the Table are made-up and taken from an example used in Alkire and Foster (2008) to represent the aggregated values of indicators within each of the domains. The values for all the indicators are not provided—just the aggregate values for the domains.

The data are provided for four imaginary respondents (A, B, C and D). The "sufficiency cutoff" values are set by experts.

Those values above the "sufficiency cutoff" are shaded.

TABLE 3.9 (Continued)

(b) Gaps between the values in (a) recorded for each respondent and the "sufficiency cutoff"

Happiness domain										
Respondent	Psychological wellbeing	Health	Education	Time use	Cultural diversity and resilience	Good governance	Community vitality	Ecological diversity and resilience	Living standards	Number of responses less than sufficiency
A	2	-1	0	1	2	1	-1	1	0	2
B	-1	0	-3	0	-3	3	-2	-2	-2	6
C	4	-4	-2	3	-3	-1	0	0	-2	5
D	-3	0	1	-4	-2	0	1	-4	-2	5
Number of responses less than sufficiency	2	2	2	1	3	1	2	2	3	18
Sum of gaps	-4	-5	-5	-4	-8	-1	-3	-6	-6	

The Figures in the Table are the gaps between the made-up data in (a) and the "sufficiency cutoff"

For the sum of gaps, only values that are less than the "sufficiency cutoff" (i.e., the negative values) are included. Those values above the "sufficiency cutoff" (shaded in the Table) are assumed to be "fully happy".

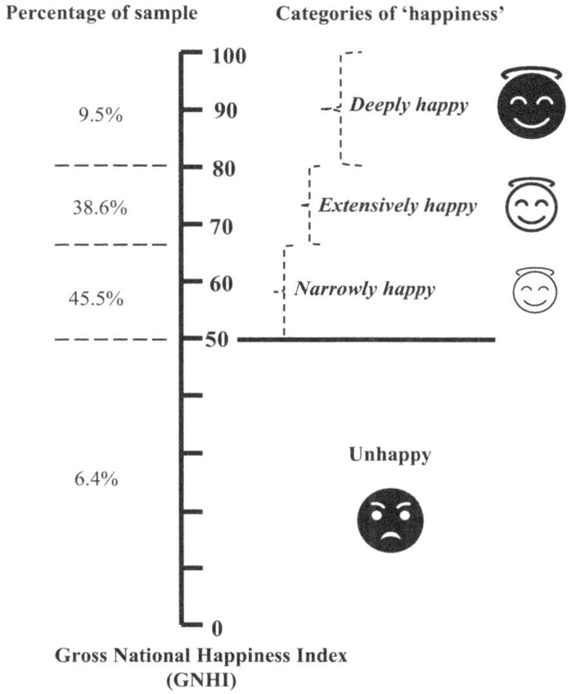

FIGURE 3.1 The categories used by the Bhutan government for values of their Gross National Happiness Index (GNHI).

Based on data available from the Oxford Poverty and Human Development Initiative (OPHI) Also shown are the percentage breakdown of a sample from 2022 into the four categories.

Economic Paradigm". After 2012, "World Happiness Reports" have been published in 2013 and every year since 2015. At the time of writing, the latest available version of the WHR was published in 2024 (Helliwell et al., 2024). The reports include rankings of countries based on their "happiness", and the principal source of the data used to construct this ranking is provided via a survey conducted by an American research company called Gallup. Gallup was founded in 1935 and is perhaps best known for regular surveys of voting intentions in some democracies, but since 2005 it has also undertaken a major international survey (The Gallup World Poll; GWP) which spans various matters of interest. As the company says on its website:

> The Gallup World Poll tracks the most important issues worldwide, such as food access, employment, leadership performance, and well-being. Since creating the World Poll in 2005, Gallup has conducted studies in more than 160 countries that include 99% of the world's adult population.
>
> *(Source: https://www.gallup.com/178667/gallup-world-poll-work.aspx)*

Interviews are conducted via telephone or face-to-face, and respondents (all over 15 years of age) are selected using a sampling frame to generate a sample that is "*probability based and nationally representative of the resident population*". The sample size per country is typically around 1,000 but can vary from 500 to 2,000. For example, the 2022 survey for the UK was undertaken using a combination of telephone and face-to-face methods between May 30 and June 27, with a sample size of 1,000 individuals. The key question in the GWP survey used to assess "happiness" is as follows (English version of the wording):

> Please imagine a ladder, with steps numbered from 0 at the bottom to 10 at the top. The top of the ladder represents the best possible life for you and the bottom of the ladder represents the worst possible life for you. On which step of the ladder would you say you personally feel you stand at this time?
>
> *(Helliwell et al., 2024; Appendix 1 page 1)*

The respondent is asked to provide a score from 0 (worst possible life) to 10 (best possible life). The device being used here is called the Cantril ladder, and is named after its creator—Albert Hadley Cantril, Jr. (1906–1969), a Princeton University psychologist, and the idea is first set out in his book published in 1965 titled "The pattern of human concerns" (Cantril, 1965). The approach taken with the ladder is referred to as "self-anchoring". Rather than try and define a common notion of "best" and "worst" across all respondents a self-anchoring scale assumes that each respondent has their own sense of what the "best" and "worst" possible lives are for them, and their scoring from 0 to 10 is relative to those self-determined "anchors". It cannot be assumed that a score given to an attribute means the same across all respondents; they are all "self-anchoring" relative to their own sense of "best" and "worst".

A second intriguing point of the question above is that it too does not actually mention "happiness" and is perhaps better referred to as "self-reported well-being". The question does not ask respondents about how happy or unhappy they are but asks respondents to score relative to a "best possible life" and "worst possible life". This seems like a more direct approach than the 9 categories of the GNHI, but can it be assumed that "best possible life" is the same as being "very happy" and vice versa? It seems possible that at least some respondents may not regard these as equivalent; after all, one can have a good "possible life" interpreted as having ownership of goods and not wanting for food etc. but not necessarily regard it as being equivalent to being "happy".

Thirdly, the question makes it clear that the assessment using the scale is very much rooted in the present (i.e., "stand at this time"). This makes sense as, of course, people's views as to where they feel they are in the scale can vary over time, and maybe even within the same day or within a single hour! But when giving a score will respondents take such "bumpiness" into account and give a kind of average "feeling" as to where they stand on the scale or will they instead be very influenced by how they feel at the very time of completing the survey? One can imagine all sorts of scenarios here, especially as respondents are being expected to answer the question in the 30 minutes to one hour period of the interview rather than having time to think.

In fairness, the authors of the WHR do make it clear that the results to the question above from the GWP is not a direct indicator of "happiness", but having said that they do then go on to link the results to what they regard as explanator factors for "happiness". For example, they explore the link between "happiness" scores and demographic factors such as sex and age. When it comes to age the link is somewhat nuanced and in the 2024 WHR, we have the following claim (emphasis by this author):

> In many but not all regions, *the young are happier than the old.* But in North America happiness has fallen so sharply for the young that they are now less happy than the old. By contrast, in the transition countries of Central and Eastern Europe, the young are much happier than the old. In Western Europe as a whole happiness is similar at all ages, while elsewhere it tends to decline over the life cycle. (with an occasional upturn for the old).
>
> *(Helliwell et al., 2024; page 5)*

Age does appear to be a major determinant of "happiness" even if the story is geographically complex. Indeed, when it comes to differences between the sexes, the picture is also complex and appears to have changed over time, especially following the COVD-19 pandemic, as well as between countries having different income levels:

> For ages 15–24, global data shows no global gender differences from 2006 until 2013. But from2014, females began reporting higher life satisfaction than males, although the gap has narrowed after the pandemic. This global gender gap masks regional differences, and is more pronounced in lower-income countries. *There are no gender differences in high-income countries.*
>
> *(Helliwell et al., 2024; page 7)*

It is also interesting to note that "happiness" inequality has also increased significantly:

> Global happiness inequality has increased by more than 20% over the past dozen years, in all regions and age groups, to an extent that differs a lot by age and by region.
>
> *(Helliwell et al., 2024; page 11)*

It seems that happiness is not evenly spread across and within countries and not even between people within the same age group.

3.7.3 What influences happiness?

As well as looking for differences between countries and gender in terms of happiness the authors of the WHR also explore how happiness may be linked to a range of variables as summarised in Table 3.10. The choice of variables was made by the WHR team, but it is certainly not hard to imagine a link between "happiness" and at least some of the variables in Table 3.10; income, social support, healthy life expectancy and freedom to make life choices can, at least superficially, be expected to have an influence on a measure of "happiness". Having a good income, social support, having a good health and having freedom to make choices in life may all be expected to improve a sense of happiness. Income is perhaps the one variable in the list that people may most likely associate with happiness; after all, money may not necessarily buy happiness, but it can certainly help! The inclusion of corruption as an "influencer" on happiness may perhaps seem less intuitive and

TABLE 3.10 Independent variables used in the World Happiness Report (2024) for regression analysis with "happiness" (Life Ladder Score) as the dependent variable (adapted from Helliwell et al., 2024).

Variable	*Notes*
Income (Log GDP per capita)	GDP per capita expressed in terms of Purchasing Power Parity (PPP) and adjusted to take inflation into account.
Social support	Determined from one of the questions in the Gallup World Poll. In this case the question is "If you were in trouble, do you have relatives or friends you can count on to help you whenever you need them, or not?" The answer to the question is either yes or no, with no 0 to 10 scoring as per the Cantril Ladder. Numerical values of "0" are given to no and "1" to yes, and an average of these is taken as the value for the country.
Healthy life expectancy at birth	This variable is based on data from the World Health Organization (WHO) Global Health Observatory data repository.
Freedom to make life choices	This variable makes use of answers to another Gallup World Poll question—"Are you satisfied or dissatisfied with your freedom to choose what you do with your life?" As per the social support question above, the allowed answers are binary (no or yes) and a national average is used in their analysis.
Generosity	As per the "social support" and "freedom to make life choices" variables, only in this case the question is "Have you donated money to a charity in the past month?"
Perceptions of corruption	Perhaps surprisingly, this variable does not make use of the CPI. Instead, it relies upon two questions asked in the Gallup World Survey: "Is corruption widespread throughout the government or not?" "Is corruption widespread within businesses or not?"

perhaps rather specific when compared to the others, and it does seem a bit odd that they did not include a wider perception of crime rather than just corruption, but as we have already seen there is certainly a case that can be made for the inclusion of corruption given its known impacts on society. The inclusion of "generosity" may perhaps be the most surprising of the "explanator" variables in Table 3.10 but the logic here is presumably that generosity can make people feel happy. So, in general, there does seem to be a solid basis for assuming that the variables listed in Table 3.10 will influence happiness. However, perhaps more intriguing than the list of explanator variables in Table 3.10 is the decision to treat happiness as the dependent variable in the analysis. Hence, it is assumed that happiness is affected by income, social support, etc., rather than the other way around. But does happiness not influence at least some of the variables in Table 3.10? Can happiness not influence life expectancy for example, or maybe even a person's propensity to make donations to charity? It is a thought.

The full details of the methodology behind the analysis are available in the WHR 2024 and do not need to be repeated here, but it does raise various points of interest that have wider relevance, and it is worth delving into these. Most of the variables listed in Table 3.10 are derived from answers to questions within the GWP survey, and that includes perceptions of corruption. Two of the variables—income and life expectancy—also appear in the Human Development Index (HDI) which itself is often interpreted to represent an assessment of quality of life or wellbeing.

The approach adopted for the analysis is called Multiple Linear Regression, and the WHR researchers have assumed that across the sample of countries the independent variables (i.e., those that are assumed to have an influence on the ladder scores equating to "happiness") are those listed in Table 3.10, but they have also included time (i.e., the year) as a variable. This makes a lot of sense as it seems reasonable to assume that ladder scores for countries will change over time. Hence the inputs into the analysis are values of each of the variables in Table 3.10 for each country and year where data are available. While it is logical, there are ramifications that arise from this decision, and to illustrate these it is necessary to first set out some of the results of the statistical analysis and one set of these is provided here as Table 3.11, which is an expanded version of the abridged version given in the WHR 2024 (Helliwell et al., 2024; Table 2.1, page 20). The complete data set used for the analysis is far too large to provide here although the reader can find them in the WHR, but the data for Afghanistan have been given in Table 3.11a as an example. The years are from 2005 to 2023 but as can be seen for Afghanistan, and the same is true for many countries, data are not complete for all years. Indeed, for some years the Ladder results are available but there are no data for some of the independent variables. In other words, it is important to note that the picture in terms of data availability is patchy, and there can be many reasons for this. In the case of Afghanistan, the recent history of conflict in that country provides an obvious yet tragic explanation. It can be so easy to forget that these numbers are about people.

TABLE 3.11 Example results of a multiple linear regression to explore how "Life Ladder" scores (a proxy for "happiness") are influenced by a range of independent variables. Based on an analysis in Helliwell et al. (2024).

(a) Example data set for Afghanistan (from the 2024 World Happiness Report)

Year	*Life Ladder*	*Log GDP per capita*	*Social support*	*Healthy life expectancy at birth*	*Freedom to make life choices*	*Generosity*	*Perceptions of corruption*
2005							
2006							
2007							
2008	3.72	7.35	0.45	50.50	0.72	0.16	0.88
2009	4.40	7.51	0.55	50.80	0.68	0.19	0.85
2010	4.76	7.61	0.54	51.10	0.60	0.12	0.71
2011	3.83	7.58	0.52	51.40	0.50	0.16	0.73
2012	3.78	7.66	0.52	51.70	0.53	0.23	0.78
2013	3.57	7.68	0.48	52.00	0.58	0.06	0.82
2014	3.13	7.67	0.53	52.30	0.51	0.10	0.87
2015	3.98	7.65	0.53	52.60	0.39	0.08	0.88
2016	4.22	7.65	0.56	52.92	0.52	0.04	0.79
2017	2.66	7.65	0.49	53.25	0.43	−0.12	0.95
2018	2.69	7.63	0.51	53.58	0.37	−0.09	0.93
2019	2.38	7.64	0.42	53.90	0.39	−0.11	0.92
2020							
2021	2.44	7.32	0.45	54.55	0.39	−0.08	0.95
2022	1.28		0.23	54.88	0.37		0.73
2023	1.45		0.37	55.20	0.23		0.74

(b) Analysis of Variance (ANOVA) results for all the data in the 2024 World Happiness Report

Source		*DF*	*Adjusted SS*	*Adjusted MS*	*F-Value*
Regression		24	2060.38	85.8492	273.6
	Log GDP/capita	1	78.54	78.545	250.32
	Social support	1	95.09	95.0875	303.04
	Healthy Life Expectancy at birth	1	22.92	22.9209	73.05
	Freedom to make life choices	1	44.97	44.9743	143.33
	Generosity	1	10.82	10.8232	34.49
	Perceptions of corruption	1	26.31	26.3071	83.84
	Year	18	17.34	0.9633	3.07
Error		2078	652.04	0.3138	
Total		2102	2712.42		

Note: Constant term has been included in the model and "Year" is treated as a categorical predictor. Adjusted $R^2 = 75.68\%$. All of the F-values in the Table are statistically significant at $P<0.001$.

DF is degrees of freedom, SS is the sums of squares and MS is the mean square.

(c) Model coefficients

Term	*Model coefficient*	*Standard error of coefficient*	*t-value*	*VIF*
Constant	−1.811	0.587	−3.09	
Log GDP/capita	0.3494	0.0221	15.82	4.35
Social support	2.563	0.147	17.41	2.22
Healthy Life Expectancy at birth	0.02771	0.00324	8.55	3.45
Freedom to make life choices	1.378	0.115	11.97	1.71
Generosity	0.4871	0.0829	5.87	1.21
Perceptions of corruption	−0.7333	0.0801	−9.16	1.45

Note: All t-values are statistically significant at $P<0.01$. The coefficients for each of the years are not presented here. VIF is the Variance Inflation Factor.

Besides missing data, a further issue with the regression analysis in WHR is the treatment of time (year) as a nominal category. Nominal categories are mutually exclusive, a single data point can only be in one of them, and there is no inherent order to the categories; a commonly cited example of nominal categories would be "male" and "female". But treating year as a nominal category is something of a fudge as strictly speaking year is an ordinal category (i.e., ordered like 1, 2, 3, 4, etc.). Year obviously has an inherent order to it which nominal variables (e.g., male and female) do not. Yet in the interpretation adopted by the WHR analysis, the year 2005 is regarded as being as different from 2023 as it is from 2006, which is not logical. One can readily expect that the ladder scores and other variables may change relatively little over the space of a year (e.g., from 2005 to 2006), but they could change a lot when compared to 2023—a gap of 17 years. An illustration for the WHR ladder scores is provided as Figure 3.2, where the graph is based on correlation coefficients (r) of ladder scores between each year and subsequent years for the entire dataset. For example, the results for 2005 are first correlated with those for 2006 (1 year later), then with 2007 (2 years later), 2008 (3 years later) and so on up to 2023 (18 years later). The process is repeated for year 2006 which is correlated with 2007 (1 year later), 2008 (2 years later) and so on. The result of this process is a matrix of correlation coefficients of ladder scores between each year and subsequent years, with in this case higher values for r equating to stronger correlation. As can be seen from Figure 3.2, the correlations between the ladder scores of each year are, as expected, closely correlated with those of the following year but less so with the year after that and so on as the years pass. This makes sense as years closer to each other are more likely to have similar results from the survey compared with years that are far apart. Therefore, Figure 3.2 makes a strong case for treating year as an ordinal category rather than a nominal one, so why have the WHR data analysts chosen to define year as a nominal type of category? No explanation is provided but one, even if somewhat unsatisfying, explanation could be that it does help make the analysis easier.

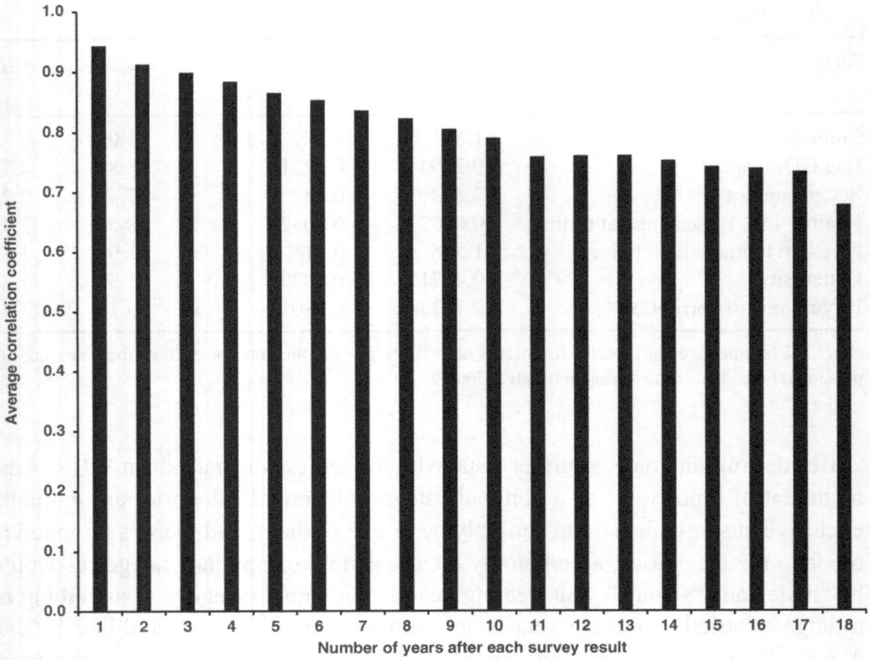

FIGURE 3.2 The level of correlation between ladder scores (assumed to equate to happiness) for one year and those of subsequent years

Note: A correlation coefficient (r) of 1.0 indicates a perfect linear association between two variables. The graph is based on a correlation coefficient matrix of ladder scores between 2005 and 2023, with one based on the correlation of survey results with the next year, two based on correlation with survey results two years later and so on. The bars are the average of the correlation coefficients. Data are available from the World Happiness Report (Helliwell et al., 2024).

Putting all these concerns over missing data and the use of ordinal categories for year to one side, the results of the multiple regression of ladder score against the independent variables listed in Table 3.10 are shown as Table 3.11b (ANOVA Table) and Table 3.11c (Table of regression coefficients). As year is treated as a nominal category, it does not have an overall coefficient in Table 3.11c and instead should have a separate coefficient for each year which would make for a long list of results. Hence, for the sake of brevity, I have not presented the regressions for each of the years. The fitted model is statistically significant, with an adjusted R^2 (coefficient of determination) of 75%, and all the independent variables in Table 3.11c provide a statistically significant contribution to the "best fit" model. Based upon these findings, the WHR analysts conclude that life ladder score (i.e., happiness) is positively influenced by income, social support, healthy life expectancy, freedom to make life choices and generosity (regression coefficients for these are positive), and negatively influenced by corruption (regression coefficient for corruption is negative). Hence, the more income people have, the more social support they perceive they have, the longer their healthy life expectancy, the greater their perceived ability to make life choices and their views on donating to charity all result in more happiness. Conversely, the more corruption they perceive to be in the country then the less their perception of happiness. All of this makes sense, but as noted above it is predicated on happiness being an outcome (dependent variable) of these variables rather than the other way around.

Another potential issue with any such analysis is that the so-called independent variables may be related to each other. For example, it is not inconceivable that income is linked to life expectancy; the more income a person may have then the greater the resources they have available to pay for better living conditions, diet and healthcare. Similarly, the greater the income, then the higher their perceived freedom to make life choices and to be generous. Significant inter-relationships between "independent" variables are what statisticians call "collinearity", and with many such variables, the term "multi-collinearity" is more applicable. This matters a great deal as the inclusion of such related variables can generate an inflated influence on the results of a regression analysis; a sort of "mutual reinforcement" which means that the regression may appear to be stronger than it perhaps is. A clue to potential inter-relationships between independent variables is provided by the rather technical sounding "Variance Inflation Factors" (VIFs) in Table 3.11c; in essence, the greater the value of the VIF, then the greater the influence of collinearity (links with other independent variables) on the regression coefficient. I will not go into the details here, but some statisticians have argued that a VIF above 5 suggests that the degree of collinearity is too high and something needs to be done, perhaps by eliminating one of the related variables, otherwise, the regression may provide inflated, and thus distorted, results (Hae Kim, 2009), while others have been more conservative and suggested a lower value of 3.3 as the cut-off for raising concern (Kock and Lynn, 2012). In Table 3.11c, none of the VIFs exceed the 5 level, but there are two which are greater than 3.3. But it does have to be acknowledged that opinion over the VIF cut-off point for concern does vary between statisticians and some have even suggested a VIF of less than 10 as being fine.

Income measured as GDP/capita (and indeed related economic indicators to GDP) tend to have a curved relationship with many social variables such as life expectancy; social variables increase with wealth but only up to a point after which they tend to flatten out. This can be seen in Figure 3.3a as "happiness" increases with wealth up to around $40,000/capita, but after that point the increase is much lower and maybe even flatlined. Looks like money can buy happiness but only up to a point; after that point, more money does not necessarily bring more happiness. Hence as seen in Table 3.10 and subsequent analyses, the income variable included in the regression is the logarithm (in this case using base e) of the GDP/capita data rather than the raw data. Taking the logarithm of a variable has the effect of reducing the spread of data points, and this can be seen in Figure 3.3b where the same GDP/capita data have been transformed using logarithms. The result is that the data appear to form more of a straight-line (i.e., are linear) than a curve.

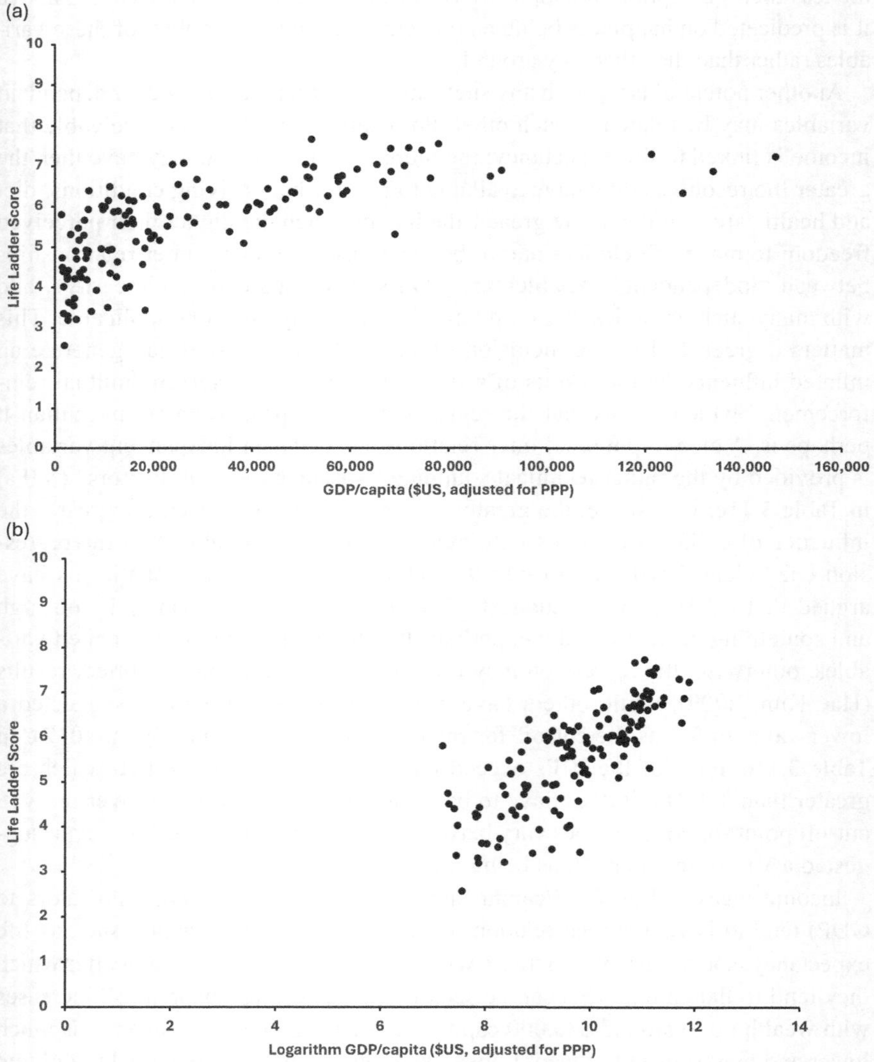

FIGURE 3.3 Relationship between Life Ladder Score (assumed to equate to happiness) and income measured as GDP/capita

Notes: (a) is based on the raw GDP/capita data (US$ adjusted for purchasing power parity) while with (b) the GDP/capita data have been transformed by talking the logarithm (base e). In both case, the year used for the analysis is 2022. Data are available from the World Happiness Report (Helliwell et al., 2024).

Education does not appear in any of the variables listed in Table 3.10 and thus is not included in the multiple regression results of Table 3.11 and that is perhaps surprising. Should we expect the level of education to have an influence on happiness? Are more educated people also happier? It may be argued that the inclusion of education takes us back to the collinearity issue noted above. After all, education level could potentially be correlated with some of the other variables such people's perception of their "Freedom to make life choices" and, of course, income. Also, while education is not included in the suit of variables in Table 3.10, it does feature in the narrative in the WHR 2024. For example:

> [S]atisfaction with schools and the education system is positively associated with life satisfaction in lower middle and low-income countries, but not in upper middle- and high-income countries.
>
> *(Helliwell et al., 2024; page 92)*

Indeed, the WHR 2024 includes a detailed analysis of the Cantil Ladder scores for India and a range of independent variables, one of which is education.

> Older adults with secondary or higher education and those of higher social castes report higher life satisfaction than counterparts without formal education and those from scheduled castes and scheduled tribes.
>
> *(Helliwell et al., 2024; page 7)*

It was estimated by the WHR researchers that the level of education contributed 7.5% towards the life satisfaction score, although this was far behind the contribution coming from satisfaction with living arrangements which was 48.2%. The link between education and life satisfaction may be a straightforward one in the sense that better education tends to give higher income, but matters may be more subtle than this:

> One potential mechanism could be that those with more education have a more diverse social network, which may translate into more diverse forms of support. Diversity in social networks also may mean interactions with different types of individuals and more diverse social activities, both of which could promote higher order processing leading to better cognitive health
>
> *(Helliwell et al., 2024; page 145)*

One final point of interest in the WHR analysis is their inclusion of "positive" and "negative" affects on life ladder scores also based on data derived from the Gallup World Survey. These are defined by them as follows (emphases provided by this author):

> *Positive affect* is defined as the average of three positive affect measures in GWP: *laugh, enjoyment and doing interesting things* in the Gallup World Poll.

> These measures are the responses to the following three questions, respectively: "Did you smile or laugh a lot yesterday?", and "Did you experience the following feelings during A LOT OF THE DAY yesterday? How about Enjoyment?", "Did you learn or do something interesting yesterday?
>
> *Negative affect* is defined as the average of three negative affect measures in GWP. They are *worry, sadness and anger*, respectively the responses to "Did you experience the following feelings during A LOT OF THE DAY yesterday? How about Worry?", "Did you experience the following feelings during A LOT OF THE DAY yesterday? How about Sadness?", and "Did you experience the following feelings during A LOT OF THE DAY yesterday? How about Anger?
>
> *(Helliwell et al., 2024; page 21)*

The WHR 2024 includes positive and negative affects as independent variables in their multiple regression analysis of Life Ladder scores, and Table 3.12 sets out the results in a more expanded form. The interpretation of this table is the same as that of Table 3.11, and all the same caveats about the way year is treated, collinearity, etc., apply equally here. Again, for the sake of brevity, the individual regressions for each year have not been included, and "Year" does not appear in the list of regression coefficients. As before, the results in Table 3.12 point to a positive

TABLE 3.12 Example results of a multiple linear regression to explore how "Life Ladder" scores (a proxy for "happiness") are influenced by a range of independent variables, including positive and negative "affects". Based on an analysis in Helliwell et al. (2024).

(a) ANOVA results for all the data in the 2024 World Happiness Report and including positive and negative "affects".

Source		*DF*	*Adjusted SS*	*Adjusted MS*	*F-Value*
Regression		26	2123.95	81.6906	288.86
	Log GDP/capita	1	92.55	92.5485	327.25
	Social support	1	43.03	43.0348	152.17
	Healthy Life Expectancy at birth	1	23.79	23.7889	84.12
	Freedom to make life choices	1	6.14	6.1435	21.72
	Generosity	1	3.86	3.8632	13.66
	Perceptions of corruption	1	24.54	24.5389	86.77
	Positive affect	1	65.9	65.8983	233.02
	Negative affect	1	0.37	0.3685	1.3 not sig
	Year	18	14.22	0.7902	2.79
Error		2070	585.41	0.2828	
Total		2096	2709.36		

Note: Constant term has been included in the model and "Year" is treated as a categorical predictor. Adjusted $R^2 = 78.12\%$. Almost all the F-values in the Table are statistically significant at $P<0.001$, with the exception of "Negative affect" which is not statistically significant.

(b) Model coefficients

Term	*Model coefficient*	*Standard Error of coefficient*	*T-Value*	*VIF*
Constant	−2.584	0.56	−4.59	
Log GDP/capita	0.3819	0.02	18.09	4.4
Social support	1.936	0.16	12.34	2.79
Healthy Life Expectancy at birth	0.02867	0	9.17	3.55
Freedom to make life choices	0.571	0.12	4.66	2.15
Generosity	0.2955	0.08	3.7	1.25
Perceptions of corruption	−0.724	0.08	−9.32	1.51
Positive affect	2.206	0.15	15.26	1.79
Negative affect	0.193	0.17	1.14 not sig.	1.55

Note: All t-values are statistically significant at $P<0.001$ with the exception of "Negative affect" which is not statistically significant. The coefficients for each of the years are not presented here.

relationship between Life Ladder scores and income, social support, Healthy Life Expectancy at birth, Freedom to make life choices and Generosity. There is also a repeat of the negative relationship between Life Ladder scores and perceptions of corruption. So far, all well and good. Life Ladder scores increased with "positive affect", which is perhaps not all that surprising as, after all, happiness would surely bring with it laughter, enjoyment and doing interesting things; these are almost "symptoms" of being happy. But what is more surprising is that Life Ladder scores were apparently not related to the "negative affect". We would probably expect that lower Life Ladder scores would be associated with feelings of worry, sadness and anger, yet the results of the regression do not point us in that direction. Why this should be is not clear. Why does the "positive affect" influence Life Ladder score but not the "negative affect"? Perhaps it is worth stepping back a little here. As you can see from the definitions of positive and negative given above, there is a difference between how these two "affects" have been determined. "Positive affect" appears to be mostly based on asking people if they did something—if they laughed and did something interesting? By way of contrast, the "negative affect" is based more on how people felt—did they feel worry, have sadness and anger? Admittedly, the "positive affect" also included the feeling of enjoyment, but even so there does seem to be a fundamental distinction between the nature of the positive and negative "affects". Asking people if they did something, as expressions of being "positive", is arguably quite different from asking people how they felt as expressions of being "negative".

3.7.4 *The cost of happiness: Happy Planet Index*

Finally, it is perhaps surprising that the WHR 2024 only mentions the term "environment" eight times in the entire report, and of those mentions, the context is typically one of "social environment" (family, friends) rather than natural environment.

Sustainability is only mentioned twice, but neither of these mentions are in the main body of the text—they only appear in the reference list! Yet happiness is well-known to be enhanced by the natural environment and people's access to green spaces (Kwon et al., 2021) and, on the more negative side, can be reduced by a perception of environmental degradation caused by people (Ferrer-i-Carbonella and Gowdy, 2007). Happiness does have a relationship with the natural environment even if this does not find its way into any of the analyses in WHR 2024 and not mentioned in the text as a potential factor. This brings us to another indicator linked to happiness that attempts to fill this gap, namely the wonderfully named Happy Planet Index (HPI) (Abdallah et al., 2024). The HPI builds upon the concept of "happiness" set out above but seeks to also include the environmental cost of "happiness" on a country-by-country basis. The HPI tries, in essence, to capture the environment cost of living a happy life. The HPI was conceived and is curated by a non-governmental organisation called the "New Economics Foundation" (NEF) having the intriguing strapline of "Economics as if the people and the planet mattered". Abdallah et al. (2024; page 4) claim that the HPI:

> [I]s based on a simple idea: The ultimate outcome for societies should be sustainable wellbeing for all. And the fundamental input is environmental resources. If we want one single indicator to get an overall sense of how we are doing, that indicator should be a measure of how much wellbeing is achieved per unit of environmental resources, in effect a measure of socio-ecological efficiency.

With such "efficiency" (output relative to input) in mind, the HPI is found for each country based on the following calculation:

$$Happy\ Planet\ Index \approx \frac{(Life\ \exp ec \tan cy\ x\ Self - reported\ wellbeing)}{Environmental\ impact}$$

Note that the equation states that HPI is assessed only approximately by the components on the right-hand side of the equation, hence the use of "≈" (meaning "approximate") rather than "=" (meaning "equals").

Life expectancy is expressed in years, and the numerator represents a "happy life expectancy", also in years, by utilising the same Cantrill Ladder results from the GWP survey used for "happiness" in the WHR. The denominator in the equation is the impact that country has on the environment. In earlier versions of the HPI, the Ecological Footprint (EF) was used as a measure of impact, while in later versions, this was switched to Carbon Footprint and thus linked more directly to global warming. Therefore, the HPI can be thought of as a sort of benefit:cost ratio with happy life expectancy being the "benefit" and environmental impact being the "cost", with higher values representing a better return in terms of "happy life expectancy" per unit impact on the environment. It is another example of a hybrid index where perception-based data are combined with non-perception-based data.

3.8 Perception indicators of safety, risk and place

3.8.1 Safety and risk

There are various indicators designed to capture perceptions of safety and risk, with the latter including both personal risk of harm as well as the risks of doing business in certain environments. An example of the former designed for comparison across countries and regions is the Safety Perceptions Index (SPI). The SPI was developed by the IEP and was first released in 2022. The index is populated using data from the WRP, a global survey designed by the Lloyd's Register Foundation and administered by Gallup; the same company that manages the GWP discussed above under happiness. The SPI:

> measures two themes, worry about harm and recent experience of serious harm, analysing them across five domains: food and water, violent crime, severe weather, mental health, and workplace safety. These themes and domains are combined into a composite score which reflects perceptions of safety by country and region.
>
> *(SPI, 2023; page 2)*

The WRP questions delve into people's perceptions and experiences of various everyday risks, as well as their greatest perceived risk and how safe they feel. Hence, the WRP is not solely about perceptions as it also asks people to self-report their personal experiences (Cumbers, 2023), although as noted earlier in this chapter and in Chapter 2, such self-reporting is potentially open to some bias. The Poll also collects information about the demographics of its respondents including their sex, income and education level. It is described as a "bottom up" barometer of perceptions and experiences of risk of everyday people around the world. There are other global surveys of risk that are based solely on perception rather than asking people to self-report experiences, and an example is the GRR published by the World Economic Forum.

In addition to the WRP and GRR, there are many other attempts to assess perceptions that people have about their own safety and what factors they think influence that feeling. This is a topic of interest to government as it can help shape their policies and indeed other decisions such as planning, licensing and law and order. The UK government, for example, runs a regular survey called the OPN designed in part to capture these perceptions and experiences of individual safety and the factors which influence it. The aim is to derive something that is more socially and geographically granular than global initiatives such as the WRP and GRR. For example, the results of the OPN survey for early 2022 led to the following conclusions amongst many others:

- People felt less safe walking alone in all settings after dark than during the day; with women feeling less safe than men in all settings after dark
- Disabled people felt less safe in all settings than non-disabled people
- More women (27%) than men (16%) reported they had experienced at least one form of harassment in the previous 12 months
- People felt unsafe "in a park or open space"; this increased from 7% to 11% during the day and from 60% to 63% after dark
- People felt less safe using public transport after dark than during the day
- Women aged 16 to 34 years felt the most unsafe of any age and sex group using public transport alone after dark

But unlike the SPI, there is no attempt to create a formal indicator of personal safety out of the OPN results, and instead, the answers to the questions are used to derive the kind of narrative-based conclusions set out above. It would not be hard to express these in terms of indicators and integrate them in some way as we have seen with many of the other perception-based indicators, but that route has not been taken here. Indicators are not always the only or even the best way forward when it comes to summarising information.

3.8.2 Perceptions of place

The final perception-based indicator from Table 3.1 I wish to cover here is that focussed on a sense of place. There is some overlap here with perceptions of risk and safety which is why I have placed them into the same section. Many of the indicators in Table 3.1 are based on surveys where people are asked for their views about a particular topic, and the results are not necessarily fed back to the respondents on an individual basis but are instead aggregated and manipulated to generate values for the indicators across samples spanning countries or perhaps demographics within countries. But another way of populating indicators would be to use an assessment where the results are immediately made available to the individual to do with as they wish, and an example is the "Wellbeing Index" that can be accessed online at www.mywellbeingindex.org. Another approach would be to collect data from people via "crowdsourcing". In this case, people may be presented with a set of questions online and asked to provide their perceptions, but there is no attempt made to select respondents using any kind of stratified sampling strategy, so anyone anywhere can provide the feedback and indeed one person can provide multiple copies and/or variants of their feedback which generates potential for exploitation (Lee, 2017). The outcome is perhaps more akin to a "customer rating" or review exercise of the type one often sees when purchasing goods and services online, but it nonetheless has its attractions. It arguably has a stronger sense of engagement as the user can enter their results and immediately see the outcome; there is no waiting for experts to check, validate and use the data to produce an indicator.

Perhaps the best-known example of a "crowdsourcing" approach to populating indicators based on perception is provided by "Numbeo"; a website housed in Serbia and founded in 2009. The Numbeo website allows people to enter values for their perceived quality of life based on the following attributes:

- Cost of living and purchasing power
- Affordability of housing
- Pollution including air, water, etc.
- Crime rates
- Health system quality
- Traffic (commute times)

The primary focus is upon cities and each of the topics listed above is further broken down into a set of questions, and some examples for the "Crime rates", "Health system quality" and "Pollution" attributes are shown in Table 3.13. These are all, of course, highly subjective in nature, and the open-access nature of the exercise allows anyone with internet access to answer multiple times which can result in a degree of "gaming" if someone is willing to put the time and effort into it (Lee, 2017).

TABLE 3.13 Numbeo online questions for gauging perception of place

Crime rates	*Health system quality*	*Pollution*
Level of Crime Perception	*Skill and competency of medical staff*	*Satisfaction with air quality*
Change in Crime Levels	*Speed in completing examinations and reports*	*Quality and accessibility of drinking water*
Daytime Safety Perception	*Equipment for modern diagnosis and treatment*	*Water pollution (general)*
Nighttime Safety Perception	*Accuracy and completeness in filling out reports*	*Satisfaction with garbage disposal*
Concerns about Mugging or Robbery	*Friendliness and courtesy of the staff*	*Perception of cleanliness and tidiness*
Concerns about Car Theft or Break-ins	*Responsiveness (waitings) in medical institutions*	*Noise pollution and light during the night in the city*
Concerns about Physical Attacks by Strangers	*Convenience of location for you*	*Availability and quality of green spaces and parks*
Concerns about Insults or Harassment in Public		*Feeling of comfort in the city due to pollution*
Concerns about Bias-Motivated Attacks		
Drug Problem Perception		
Property Crime Severity		
Violent Crime Severity		

3.9 Populating perception-based indicators

In many cases, indicators can be populated with data that has already been collected by someone else, and examples of this such as the CPI have already been discussed. But there are times when new data are required, and even the most well-conceived indicator will be useless if there are no data of a good enough quality to populate it. There are many challenges here in terms of sampling frames, how often the data are to be collected, the geographical coverage, who collects the data and so on, and all this activity comes at a cost. Cutting corners to save money with little in the way of checks and balances could end up generating data of limited quality, and this can be expressed via the often-quoted maxim of "garbage in then garbage out", often attributed to George Fuechsel, an IBM computer programmer, in 1957. Putting bad quality data into an indicator is not necessarily going to help in terms of guiding interventions likely to succeed. Incidentally, the need for data of good quality to populate indicators applies whether the data have come from surveys or other methods, including measuring instruments. If a city has a set of instruments which measure aspects of air quality (NOx, SOx, carbon monoxide, particulates, etc.), with those data being used as the basis for an indicator of air quality, the value of the indicator will still depend on key decisions over what is being measured, their sensitivity, where the instruments are located and at what times of the day they are designed to work and so on. There is still potential for error here as, for example, the instruments may miss key pollutants or be in the wrong places; hence, even if the measurements are highly precise and reliable, the resulting indicator may provide only a partial picture of the quality of the air and may miss much of importance. This is not the fault of the machines, but decisions made by people about what the machines should measure, where and when.

Data for populating perception-based indicators can be obtained in various ways, including the use of participatory methods, surveys (such as the Gallup World Survey) and online questionnaires as with the "Numbeo" example. A widely used approach in the population of indicators based on capturing people's attitudes and perception via questionnaires is the Likert Scale (Likert, 1932; Likert et al., 1993). It is named after its originator, Rensis Likert, who undertook PhD research at Columbia University in the USA during the early 1930s on the quantitative assessment of attitudes. Likert sought to build upon work undertaken a few years earlier (late 1920s) by Louis Thurstone (Thurstone, 1927, 1928a, 1928b). Indeed, Thurstone's scale is often regarded as the first attempt to generate a formal means for quantifying people's attitudes and is a multi-stage process.

Step 1: Identify the aim of the research; what is it that we are trying to identify attitudes about?

Step 2: A set of potential "agree/disagree" choice statements are formulated which are designed to explore attitudes to what it is we are interested in. An example is provided by Huria et al. (2023) in their exploration of attitudes towards organic farming in India. The list of initial statements associated with organic farming is presented in Table 3.14, comprising 45 statements in total all created by the researchers. The list in Table 3.14 is certainly a long one, covering all sorts of aspects of organic farming, but there is no set or accepted notion of how long such a list should be.

TABLE 3.14 List of statements linked to organic farming in Ladakh, India (table adapted from Huria et al., 2023)

Statement number	*Statement*
1	Organic farming is an effective way of conserving natural resources.
2	A farmer can never obtain higher yields without using agrochemicals.
3	Organic farming will fetch profit over a long time.
4	Adoption of organic farming is a gamble.
5	I would like to recommend every farmer to be an organic farmer.
6	It will be better for farmers to focus on the judicious use of agrochemicals rather than organic farming.
7	Organic farming can prove to be helpful in improving the livelihood of the farmers.
8	Organic farming is not practically possible for farmers.
9	Organic farming is the need of the hour.
10	Sole dependence on Organic farming will create a shortage of food.
11	Organic farming provides healthy food to the family.
12	Produce from chemical farming is as good as organic produce.
13	Farmers should prefer organic farming over chemical farming.
14	Organic farming does not insure good income.
15	Organic farming is economical than chemical farming.
16	Adoption of organic farming at small scale is worthless.
17	Organic farming strengthens the use of indigenous knowledge.
18	Organic farming is beneficial for only large farmers.
19	Organic manures applied in farm does not affect consumers' health.
20	Adoption of organic farming is not possible due to labour shortage in Ladakh.
21	If I do organic farming, then my family members will have better health condition.
22	I will have the problem in sourcing and purchasing of organic inputs.
23	Organic farming is necessary to make Ladakh a "Carbon Neutral Region".
24	Chemical herbicides are more suitable to control weeds than organic methods of weed control.
25	Organic farming has the potential to revive the shrinking biodiversity in Ladakh.
26	Scientists have insufficient knowledge and training of organic farming practices.
27	Organic farming can restore the threatened traditional farming culture in Ladakh.

(*Continued*)

TABLE 3.14 (Continued)

Statement number	*Statement*
28	Supervision of organic farming activities will be very difficult.
29	Organic produce fetches more price in the market as compared to non-organic produce.
30	Getting organic certification is a very complex process.
31	Organic farms and products in Ladakh can also be a source of attraction for tourists.
32	Organic fertilisers are costlier than chemical fertilisers.
33	Organic products from Ladakh have the potential to fetch premium prices across the globe.
34	Organic farming can restore and maintain the balance in soil health.
35	Organic farming is not beneficial for farmers.
36	Organic farming has less risk of environmental pollution.
37	There is lack of knowledge about organic measures of nutrient and pest management among farmers.
38	Organic farming can make Ladakh self-reliant in terms of food sufficiency.
39	It is difficult to persuade every farmer in a village to adopt organic farming.
40	The profits earned from organic farming can prevent unnecessary rural-urban migration.
41	Organic farming can only be successful if all the farmers of a village or proximal cluster adopt it.
42	Ladakh has the potential to become the brand ambassador of Indian Organic Produce.
43	The results of organic measures of pest control are not convincing.
44	Research organisations should develop and propagate organic package and practices appropriate for Ladakh.
45	There is a lack of marketing channels for organic produce in Ladakh.

Step 3: Perhaps understandably given the length of the list in Table 3.14, the next step involves whittling down the list of statements to those that are deemed to be the most effective at capturing people's views on the topic. This can be done in various ways, and perhaps the simplest is for the research team to do it themselves although it seems odd for the researchers to first generate a (very) long-list of statements and then immediately whittle them down! Why not begin with the short-list? There would also seem to be much potential for bias on the part of the researchers.

An alternative approach, and indeed the one championed by Thurstone in his pioneering work, is to allow an independent group of "judges" (often called the "judging" or "consensus" group) outside of the research team to undertake this short-listing process instead. Statements such as those in Table 3.14 can be ranked or scored by the judges in terms of their effectiveness, and Thurstone suggested the use of a scale from 1 to 11 where 1 indicates "least favourable" (or "least effective") and 11 is "most favourable" (i.e., "most effective"). It is important to note that the judges are not being asked to score the statements in terms of whether they agree with them or not but whether they believe that the statement would be effective at capturing people's views on the topic. Hence, with statement 1 in Table 3.14, a judge may agree or disagree as to whether organic farming is an effective way of conserving natural resources, but that is irrelevant. What matters is whether they regard the statement as an effective one for teasing out people's views on the topic. That may appear to be a rather subtle distinction, but it is nonetheless a critical one, and the judges do need to be briefed by the researchers on what it is they are scoring. Indeed, it would also seem appropriate for the judges to be cognisant of the topic area at the heart of the research so they can make the kind of judgement required. There is no specification as to how large the judging group should be, and neither is there a specified scale for the judges to apply. While Thurstone suggested a scale of 1 to 11 for assessing how effective the statements are when it comes to capturing attitude, there is nothing sacrosanct about that range. One could equally use another range. While there remains the potential for bias, the judges are external and presumably independent to the research team and thus can provide a valuable check on what is important.

An example short list of statements, adapted from the complete list in Table 3.14 for the Huria et al. (2023) study on organic farming, is provided in Table 3.15. This shortlist was derived from the views of a group of 30 judges. Also included in the table are the median scores for the statements as provided by the group of 30 judges using a scoring range of 1 to 9, with 1 being extremely unfavourable, 5 neutral and 9 being extremely favourable. Thus, for every statement, there will be a total of 30 scores; assuming, of course, that each judge scores each statement. For each statement, it is then possible to produce a mean or median score of how effective that statement is deemed to be when it comes to capturing attitude towards organic farming. The median, of mid-point, is more appropriate than the mean as there is less chance of a distortion effect coming from a few extreme scores for a question. The higher the median for a statement then the greater is its perceived effectiveness for capturing an attitude, and vice versa.

However, simply taking the statements that score the highest and using them to capture attitude does miss out the degree to which there may or may not be agreement between the judges. One statement may have a high median score but may also have attracted a very wide distribution of scores from the judges, while another statement may have a similar or lower median score but be more uniform across the judges in terms of score. Hence, a degree of agreement in scoring a statement would perhaps be just as important as its favourability as captured by the median score, and that helps to explain why some of the scale values (median scores) in Table 3.15 are relatively low. There are various ways in which variation or spread between a group of scores can be measured, and a widely used one is the range—the gap between the highest and lowest value. Therefore, for each of the

TABLE 3.15 Final set of 16 statements selected from those in Table 3.14 by a group of 30 judges (Table adapted from Huria et al., 2023)

Statement number	*Statement linked to organic farming*	*Scale value*
11	Organic farming provides healthy food to the family	9
31	Organic farms and products in Ladakh can also be a source of attraction for tourists	9
29	Organic produce fetches more price in the market as compared to non-organic produce	9
1	Organic farming is an effective way of conserving natural resources	9
13	Farmers should prefer organic farming over chemical farming	9
34	Organic farming can restore and maintain the balance in soil health	9
30	Getting organic certification is a very complex process	8
43	The results of organic measures of pest control are not convincing	8
44	Research organisations should develop and propagate organic packages and practices appropriate for Ladakh	7
22	I will have the problem in sourcing and purchasing of organic inputs	7
24	Chemical herbicides are more suiTable to control weeds than organic methods of weed control	6
38	Organic farming can make Ladakh self-reliant in terms of food sufficiency	6
4	Adoption of organic farming is a gamble	5
6	It will be better for farmers to focus on judicious use of agrochemicals rather than organic farming	5
18	Organic farming is beneficial for only large farmers	3
12	Produce from chemical farming is as good as organic produce	2

statements, we can calculate both the median and the range. However, rather than use the range, Thurstone recommended the use of what is called the Interquartile Range (IQR); the range based on the two middle quartiles either side of the median rather than the whole dataset. The use of the IQR does reduce the chance of a distortion that may come from including outliers (extreme low and high values) as this can be an issue with the use of the range based on all the data. For example, a median score of 4 might be obtained where the majority (28 out of the 30 judges) of the scores are between 3 and 5, as well as one score each of 1 and 9. The range based on all the scores would be 8, but this is highly misleading. For the vast bulk of the scores, the range is only 2, and this is far more representative of the scores. Therefore, the IRQ for each question is used as a measure of agreement between the judges; lower interquartile ranges equate to better agreement, while higher interquartile ranges equate to less agreement and more ambiguity over the effectiveness of the statement.

Based upon the median and IQR of scores, a selection is made of the statements to be used in an attitudes survey to be administered to a sample of the population. In effect, the scores provided by the judges are used to select statements that they think would be most effective (most favourable) at determining attitudes towards the subject that has been chosen. While the median and IQR values provide an apparently objective assessment of the favourability of the statements, there is inevitably still much subjectivity here, and to some extent, the medians and IQRs need to be balanced against each other. We could, for example, use the IQR as the "decider" when it comes to selecting statements having similar median scores. Alternatively, some statements could perhaps be excluded even though their median score is high if the IQR is also very large, suggesting a significant level of disagreement and thus ambiguity between the judges. There are lots of subjective judgements being made here, and it helps to be as transparent as possible over how the decisions to accept and reject statements are being made.

Step 4: The shortlisted statements, such as those in Table 3.15, are used for designing a questionnaire to capture attitudes towards the topic. In the questionnaire, the respondents are invited to either agree or disagree with the statement rather than give them a score. We can, of course, add up all those who agree and disagree with each of the statements and use that as a simple assessment of attitude, but the process can also be taken a step further. Any response of "disagree" can receive a score of 0, while any responses of "agree" could be given the median score for that statement as determined by the judges in step 3. For example, someone disagreeing with statement 11 (Organic farming provides healthy food to the family) in Table 3.15 could generate a value of zero, but someone agreeing with this statement could generate a value of 9 which is the median score for effectiveness received by the judges. In effect, the use of the judges' median score for the statements provides a degree of "weighting" of the "agree" responses. Once all the values have been allocated to the responses provided by respondents, the median of these values can be interpreted as how strongly an individual respondent supports the subject matter (organic farming), with greater values implying that there is more support.

While the Thurstone scale remains highly regarded and still used as shown by the example above on organic farming in Ladakh, a significant disadvantage is the laborious nature of the steps involved. Coming up with potential statements, identifying a group of experts and asking them to assess the effectiveness of this list, followed by a process of selection of statements to be included in a survey of respondents and finally the analysis of the agree/disagree data to generate "attitudes" is somewhat convoluted (Willits et al., 2016). Likert was aware of this and sought to greatly simplify the Thurstone process for determining attitudes while maintaining rigour (Willits et al., 2016). So rather than have a multi-stage process based on scoring by a group of judges as the basis for selecting statements which are included in the agree/disagree based survey, Likert allowed survey respondents to apply a score themselves based on an ordinal scale spanning levels of agreement and disagreement with a statement. In effect, he did away with the judging group of

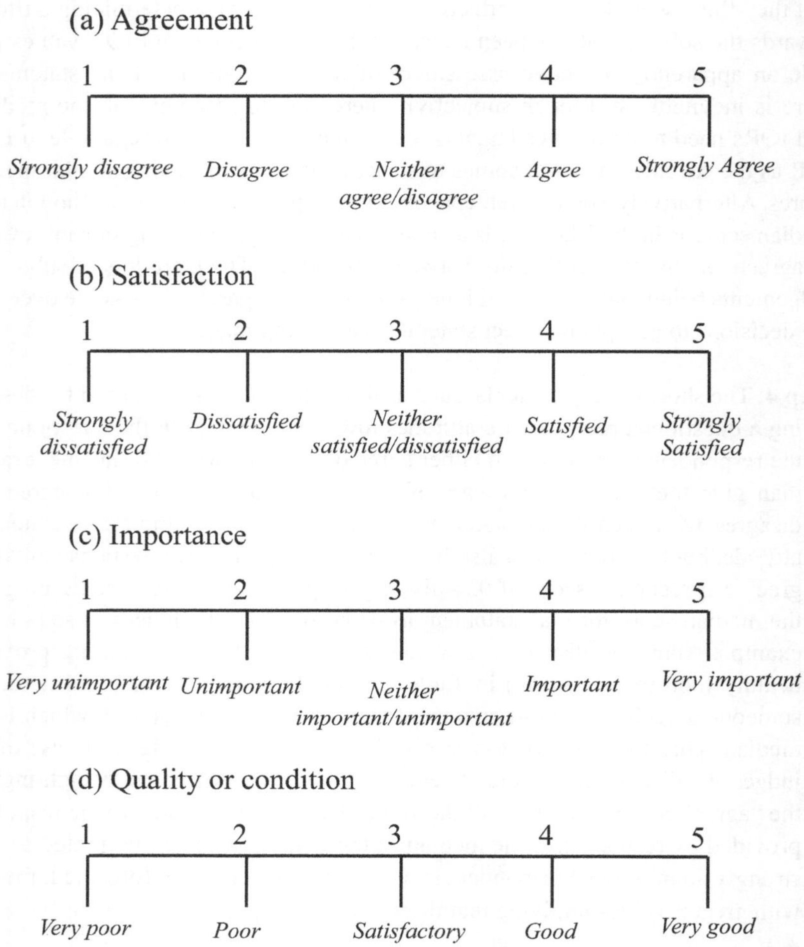

FIGURE 3.4 Examples of Likert scales (after Vagias, 2006)

All of these examples are based on Likert scales from 1 to 5.

experts. With the Likert scale approach rather than being asked to agree/disagree (a binary choice), respondents are offered an ordinal scale from 1 to 5 or 1 to 7 with "highly agree" at one end and "highly disagree" at the other. Respondents are not being forced to say agree/disagree but are given the option to "mildly" agree or disagree. Four examples of Likert-type scales, all based on a scale running from 1 to 5, are provided as Figure 3.4 and further examples can be found in Vagias (2006). The first of the examples (Figure 3.4a) is the one that more closely matches the agree/disagree choice in the Thurstone approach with respondents allowed to select from five choices including a neutral position (nether agree/disagree) in the middle of the scale. The other three examples in Figure 3.4 are, in essence, variations on the theme with Figure 3.4d perhaps being the one that most readers would be familiar with as this is often the format used by those selling goods and services as customer feedback.

The number of respondents providing each of the scores can be displayed as bar graphs, and frequency distributions for these answers may take all kinds of shapes as shown in Figure 3.5, including skewed (positive and negative) as respondents

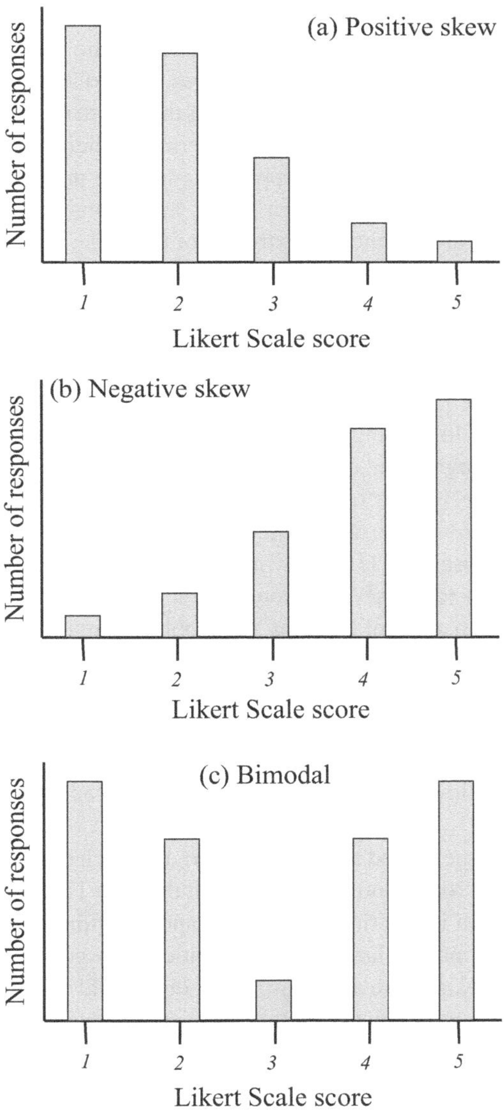

FIGURE 3.5 Some potential frequency distributions of Likert scale scores

tend to give higher and lower scores or maybe even bimodal where respondents tend to have polarised views at either end of the axis with few in the middle. Also, an important assumption with this scale is that it is linear in terms of attitude, but this may not necessarily be the case, and the interval (or gap) between "strongly agree" and "agree" may not necessarily be equal to the gap between "agree" and "nether agree/disagree"). A respondent may have some difficulty in their mind about whether they "agree" or "strongly agree" and thus be inclined to select either of these two (i.e., the gap between them is small), but they may have a much stronger sense that they "agree" with a statement rather than being undecided. The same may apply at the other end of the scale. Hence, the neat equally spaced gaps typically shown for the Likert scales in Figures 3.4 and 3.5 may be inappropriate, and given the skewed, bimodal, etc., nature of these distributions the use of means and standard deviations, while tempting in terms of their relative familiarity and simplicity, is also likely to be inappropriate unless the data just happen to have a more normal distribution (Edmondson, 2005; Sullivan and Artino, 2013). Indeed, if there is clustering around the two extremes of the scale, then the average is likely to be somewhere in the middle and imply that the respondents' "neither agree/disagree" response which is hardly representative (Sullivan and Artino, 2013).

One final point with the Likert and Thurstone approaches is how to address those who simply say they do not know or feel they cannot or do not want to give any of the responses on the Likert scale. How should their responses be addressed? One approach would be to assume that they come under the middle category on the scale (e.g., neither agree nor disagree). Therefore, the response of "don't know" or "don't want to answer" is treated as "indifference" and thus neutral, but is this really the case? Is someone saying they "don't know" really the same as saying they neither agree/disagree? Is it really indifference or being neutral? A second way of approaching such responses is to consider them as a new category on the Likert scale and give them a score of zero, but the problem here is that this category could span a range of very different responses from "do not know" to "do not care" and maybe even a fear of responding. Another way of handling such responses is to consider them as a category but not to give them a score. In this case, the responses are treated as "missing values" and not included in any statistical analysis, yet they would be worth following up all the same to understand why respondents have answered this way.

Despite all the issues noted above, the Likert approach does seem to have a more "user friendly" but "rough and ready" feel to it than the Thurstone approach (Drasgow et al., 2010) and is less time-consuming and laborious (Edwards and Kenney, 1946). The Likert approach has also had its critics (Drasgow et al., 2010) but given that they are both designed to assess attitudes, one would expect that the Thurstone and Likert methods would generate much the same results for a given sample of respondents, and indeed, some argue that is the case (Drasgow et al., 2010). While the Likert approach ostensibly does away with the need for a pre-selection stage with "judges", it may still be a wise move to seek some external (to the researcher) advice for the selection of statements to which respondents are being asked to provide a score. Nonetheless:

> The use of Likert Scales and Likert–type items have served the research community well though the years. Despite criticisms levelled against their use, analysis using Likert scales and Likert type items has contributed to advancements of knowledge in sociology, psychology, political science, biology, economics, marketing, medicine, and other fields.
>
> *(Willits et al., 2016; page 135)*

Even so, they are just one tool in the toolbox and there are other approaches to exploring attitudes that may be applicable and enlightening (Young, 2017). For all their usefulness, care needs to be taken to avoid an over-emphasis on such scales to assess perception.

As can be imagined, both the approaches of Thurstone and Likert can be applied to aid the process of selecting indicators and indeed to give indicators a value; to populate them. For example, all the statements in Tables 3.14 and 3.15 that relate to organic farming can be used as the basis to identify an appropriate indicator. Hence, a "long list" of potential indicators could be whittled down to a more manageable short list, akin to the progression from Table 3.14 to the shortlist in Table 3.15. Indeed, some of the statements in Table 3.14 do appear to be perception-based. For example, the statement that "Adoption of organic farming is a gamble" may well be an important one but assessing "gamble" in any objective sense could be challenging and is likely to vary between individuals. An indicator designed to address this statement could well involve a score from 1 to 5 depending on how much of a gamble the adoption of organic farming is perceived to be, with 5 being "very high" and 1 being "very low" (3 would be neither high nor low). Indeed, the use of a scoring or ranking process to select indicators is certainly one way of approaching the challenge of arriving at a set of agreed indicators. It has a "feel" of objectivity even if the scoring is based on judgement and that in turn can be based on many factors including bias. Nonetheless, if the selection is made by a wider group of experts or those who are at least familiar with the issue or topic, then it does lend the selection a degree of credibility. But the use of scoring along with medians and IQRs to arrive at a set of "best bets" is just one way of achieving this, and there are other ways of arriving at consensus using participatory methods, some of which will be covered in Chapter 4.

3.10 Summary

What does this foray into the world of perception-based indicators tell us? First, it is important to note that there are many of them; they are by no means a small subset of the diversity we see in the indicator multiverse. Second, some of them have been around for a long time. Indeed, people have been running surveys designed to explore attitudes, views, opinions and perceptions for decades, and although not all these surveys would necessarily have been geared towards populating indicators, they do nonetheless show that this activity is regarded as important. People's views/attitudes/perceptions do matter, and it is not hard to see how these data would have provided the foundation for indicators. Third, there are intriguing dimensions that rest behind all of this regarding definition and methods of assessment. These may not necessarily be the most alluring of dimensions, but they are important as they are vital to take the notion of a perception-based indicator from theory to practice. In this regard, perception-based indicators are perhaps no different from their non-perception-based relatives. Whenever one wishes to establish a new indicator, it is still necessary to define what it is trying to assess as well as how often it needs to be populated by whom and how often. Those are fundamental concerns for any indicator, and there are widely reported frameworks such as SMART, CREAM and SPICED designed to capture some of these concerns. But with perception-based indicators, it is often necessary to use a "translation tool" to allow for something based on subjectivity to enter the world of numbers. Hence, the prevalence and popularity of approaches based on the methodology of Thurstone and Likert scales and other devices, even if these raise other concerns.

Allied to the use of scores and counts to capture perceptions is the way in which the data are collected, and this brings us to sampling and assessment of variability between respondents. Assessing variability in data and making use of that to draw out valid conclusions is an ever-present concern within statistical analysis, of course, and is by no means limited to indicators. Very often the variability within a dataset is hidden, and values are presented in such a way that the reader may perhaps not realise that there are many assumptions at play. As noted in a previous chapter, even a measure such as GDP/capita, which seems so objective is based on a degree of subjectivity about what should be included and how it is measured. Change those assumptions and methods and the value of the GDP will change. Hence, there is a whole family of related measures such as the GNP and GNI, which differ from the GDP, and indeed many have argued that GDP should be replaced with measures such as the GPI and ISEW.

Sampling can also be an issue and obviously needs to be undertaken carefully in order to ensure the sample is representative of a wider population. Alternatives, such as the use of crowd-sourcing that allows anyone to input their scores on a topic may have all the feel of being reviews rather than a scientifically designed survey and indeed is potentially open to manipulation (Lee, 2017). Nonetheless, such crowd-sourcing approaches do have an appeal. They can be a great way to engage many people who can feel they have a voice and thus be empowering in a way that scientific-based surveys cannot match. Traditional surveys can have large sample sizes, comprising a few hundreds or thousands of people, but respondents may never see the results of their engagement, while crowd-sourcing via online tools can engage so many more, and those people can instantly see the results of their engagement.

One of the intriguing aspects of perception-based indicators is the variety of topics they seem to span as evidenced by the selection of examples in Table 3.1. Just about anything, one can consider as being part of sustainability are in here plus many others such as indicators of beauty and art appraisal. Also, some of the indicator "frameworks" set out in Table 3.1 are not called indicators per se. The suggested frameworks for art appraisal, for example, do not call the "characteristics" to be looked for "indicators" even though they look like indicators and have a very similar feel. The same applies to "happiness". The curators of the WHR that makes use of data collected by the GWP survey stress that they do not publish an "index of happiness", which indeed they do not although it is interesting that the same data are used in the HPI. The curators of the WHR make use of data from some questions as a de facto measure of happiness for their various regression analyses but they do not claim it to be an indicator. Therefore, with these perception-based assessments, we can be in something of a grey and undefined territory where there are claimed indicators (e.g., the CPI) and entities that look like indicators and seem to be used like indicators but are not claimed to be indicators. There are also the hybrids of perception-based measures with other non-perception-based data such as that in the HPI. The landscape of perception-based indicators is indeed a fascinating and highly complex one.

3.11 Food for thought

1. What other perception-based indicators are there and how do they have points of similarity and difference to those in Table 3.1?
2. What wider lessons of relevance to sustainability can be learnt from the use of indicators designed to address beauty and appearance?
3. Critically compare and contrast the three approaches taken to assessing happiness set out in this chapter.
4. Should quality of life and well-being only be assessed using perception-based indicators?

References

Abdallah S, Hoffman A and Akenji L (2024). *The 2024 Happy Planet Index*. Hot or Cool Institute, Berlin

Abildgren K and Kuchler A (2019). Revisiting the Inflation Perception Conundrum, Danmarks Nationalbank Working Papers, No. 144, Danmarks Nationalbank, Copenhagen

Abildgren K and Kuchler A (2021). Revisiting the Inflation Perception Conundrum. *Journal of Macroeconomics* 67, 103264

Alkire S and Foster J (2007, revised in 2008). Counting and Multidimensional Poverty Measurement, OPHI Working Paper 07, Oxford Poverty and Human Development Initiative (OPHI), University of Oxford

Amundi (2023). *Perceptions of Equal Opportunity*. Available at: https://about.amundi.com/article/exploring-perceptions-equal-opportunity-0

Barrett T (1994). *Criticizing Art: Understanding the Contemporary*. Mayfield Publishing Company, Mountain View, California

Baumann H (2020). The Corruption Perception Index and the Political Economy of Governing at a Distance. *International Relations* 34(4), 504–523

Bello y Villarino J-M (2021). Measuring Corruption: A Critical Analysis of the Existing Datasets and Their Suitability for Diachronic Transnational Research. *Social Indicators Research* 157, 709–747

Bhutan (2008). Constitution of Bhutan: Article 9. Available on the Constitute website: https://www.constituteproject.org/constitution/Bhutan_2008?lang=en

Bhutan's Gross National Happiness Index 2008. OPHI Research in Progress Series 2008, Oxford Department of International Development, University of Oxford, Oxford, UK

Blei DM, Ng AY and Jordan MI (2003). Latent Dirichlet Allocation. *Journal of Machine Learning Research* 3, 993–1022

Bollen KA (1986). Political Rights and Political Liberties in Nations: An Evaluation of Human Rights Measures, 1950 to 1984. *Human Rights Quarterly* 8(4), 567–591

Bollen KA and Paxton, P (2000). Subjective Measures of Liberal Democracy. *Comparative Political Studies* 33(1), 58–86

Brachinger HW (2008). A New Index of Perceived Inflation: Assumptions, Method, and Application to Germany. *Journal of Economic Psychology* 29, 433–457

Brooks JS (2013). Avoiding the Limits to Growth: Gross National Happiness in Bhutan as a Model for Sustainable Development. *Sustainability* 5, 3640–3664. https://doi.org/10.3390/su5093640

Cantril H (1965). *The Pattern of Human Concerns*. Rutgers University Press, New Brunswick, NJ

Carney JD (1994). A Historical Theory of Art Criticism. *Journal of Aesthetic Education* 28(1), 13–29

Clements RD (1979). The Inductive Method of Teaching Visual Art Criticism. *Journal of Aesthetic Education* 13(3), 67–78

CNBC (2014). *How the Corporate Perception Indicator Was Created*. Available at: www.cnbc.com/2014/09/21/how-the-corporate-perception-indicator-was-created.html

Cumbers S (2023). *What Can We Learn from Different Perspectives on Risk?* Available at: www.lrfoundation.org.uk/news/what-can-we-learn-from-different-perspectives-on-risk?gad_source=1&

Darda KM and Cross ES (2022). The Role of Expertise and Culture in Visual Art Appreciation. *Scientific Reports* 12, 10666

Davies H (2004). Dossier of Dirty Deeds. The Times Higher Educational Supplement, page 23, July 30, 2004

Doig A and McIvor S (1999). Corruption and Its Control in the Developmental Context: An Analysis and Selective Review of the Literature. *Third World Quarterly* 20(3), 657–676

Drasgow F, Chernyshenko OS and Stark S (2010). 75 Years After Likert: Thurstone Was Right! *Industrial and Organizational Psychology* 3, 465–476

Edmondson D (2005). Likert Scales: A History. CHARM Volume 12. The Future of Marketing's Past

Edwards AL and Kenney KC (1946). A Comparison of the Thurstone and Likert Techniques of Attitude Scale Construction. *Journal of Applied Psychology* 30(1), 72–83

Ferrer-i-Carbonella A and Gowdy JM (2007). Environmental Degradation and Happiness. *Ecological Economics* 60, 509–516

Freedom House (2024). *Countries and Territories*. Available at: https://freedomhouse.org/countries/freedom-world/scores

Giannone D (2010) Political and Ideological Aspects in the Measurement of Democracy: The Freedom House Case. *Democratization* 17(1), 68–97

Gilman SC (2018). To Understand and to Misunderstand How Corruption Is Measured: Academic Research and the Corruption Perception Index. *Public Integrity* 20, S74–S88

Goudie AW and Stasavage D (1998). A Framework for the Analysis of Corruption. *Crime Law and Social Change* 29(2–3), 113–159

Gutierrez L (2023). Art and Creativity: Essential Components of Sustainability. *Sustainable Earth*. Article available at: sustainable-earth.org/art-and-creativity/#:~:text=Overall%2C%20art%20and%20creativity%20are%20vital%20components%20of,actions%2C%20and%20develop%20new%20solutions%20to%20our%20challenges

Hae Kim J (2009). Multicollinearity and Misleading Statistical Results. *Korean Journal of Anesthesiology* 72(6), 558–569

Harvey AC and Proietti T (2005). *Readings in Unobserved Components Models. Advanced Texts in Econometrics*. Oxford University Press, Oxford

Helliwell JF, Layard R, Sachs JD, De Neve J-E, Aknin LB and Wang S (eds) (2024). World Happiness Report 2024. University of Oxford, Wellbeing Research Centre

Heras M, Galafassi D, Oteros-Rozas E, Ravera F, Berraquero-Díaz L and Ruiz-Mallén I (2021). Realising Potentials for Arts-Based Sustainability Science. *Sustainability Science* 16, 1875–1889

Hisamatsu Y (2003). Does Foreign Demand Affect Corruption? *Applied Economics Letters* 10(1), 1–2

Huria A, Angchok D, Tamchos T and Chaurasia OP (2023). A Scale to Measure the Attitude of Farmers Towards Organic Farming in Ladakh. *Indian Research Journal of Extension Education* 23(3), 98–103

Joshi Y and Rahman Z (2015). Factors Affecting Green Purchase Behaviour and Future Research Directions. *International Strategic Management Review* 3(1–2), 128–143

Kaufmann D and Kraay A (2024). *The Worldwide Governance Indicators. Methodology and 2024 Update*. Policy Research Working Paper 10952. World Bank, Prosperity Practice Group, Washington, DC. Available at: https://documents1.worldbank.org/curated/en/099005210162424110/pdf/IDU-7c6f0b9e-f0c2-4b1d-b30d-76c4644af69e.pdf

Kock N and Lynn GS (2012). Lateral Collinearity and Misleading Results in Variance-Based SEM: An Illustration and Recommendations. *Journal of the Association for Information Systems* 13(7), 546–580

Kumar VA, Hulst van der RRWJ, Shome DD, Booi DI, Tandel HA and Mhatre PD (2023). Multidimensional Indicators as Enablers on Perception of Facial Beauty and Appearance among Indian Population: A Descriptive Cross-Sectional Study. *Journal of Cutaneous and Aesthetic Surgery* 16(4), 292–299

Kwon O-H, Hong I, Yang J, Wohn DY, Jung W-S and Cha M (2021). Urban Green Space and Happiness in Developed Countries. *EPJ Data Science* 10(28), 1–13

Lambsdorff JG (1999). Corruption in Empirical Research—A Review. Transparency International Working Paper

Lee R (2017). How One Swede Made a City the World's 'Most Dangerous' to Expose Fake Stats. thelocal.se. Retrieved from February 7, 2022

Likert R (1932). A Technique for the Measurement of Attitudes. *Archives of Psychology* 22(140), 1–55

Likert R, Roslow S and Murphy G (1993). A Simple and Reliable Method of Scoring the Thurstone Attitude Scales. *Personnel Psychology* 46(3), 689–690

Maniatis P (2016). Investigating Factors Influencing Consumer Decision-Making While Choosing Green Products. *Journal of Cleaner Production* 132, 215–228

Manning M, Fleming CM and Wong GTW (2022). What Matters More, Perceived or Real Crime? *Social Indicators Research* 163, 1221–1248

Monaco E (2016). Notes on Bhutan's Gross National Happiness and Its Measurement. *Journal of Management and Development Studies* 27, 1–15

Morse S (2023). *Ranking Nations: The Value of Indicators and Indices*. Edward Elgar, Cheltenham, Gloucestershire

Müller H and Hornig N (2020). Expecting the Unexpected: A new Uncertainty Perception Indicator (UPI)—Concept and First Results, DoCMA Working Paper, No. 1-2020, TU Dortmund University, Dortmund Center for Data-based Media Analysis (DoCMA), Dortmund

Müller H, Rieger J and Hornig N (2021). Riders on the Storm—The Uncertainty Perception Indicator (UPI) in Q1 2021. DoCMA Working Paper #7, July 2021

Numbeo. Available at: www.numbeo.com/quality-of-life/

OECD (2023). *Hard Work, Privilege or Luck? Exploring People's Views of What Matters Most to Get Ahead in Life*. OECD Centre on Well-being, Inclusion, Sustainability and Equal Opportunity (WISE), OECD. Available at: www.oecd.org/en/about/directorates/centre-on-well-being-inclusion-sustainability-and-equal-opportunity.html

O'Neill R, Ralph J and Smith PA (2017). *Inflation. History and Measurement*. Palgrave Macmillan Cham

Oxford Poverty and Human Development Initiative (OPHI). *Gross National Happiness*. Available at: https://ophi.org.uk/gross-national-happiness

Safety Perceptions Index 2023. *Understanding the Impact of Risk around the World*. Lloyds Register Foundation, London. Available at: www.lrfoundation.org.uk

Seide M (2024). *What Scientists Have to Say about Facial Beauty*. Available at: www.verywellmind.com/what-defines-beautiful-according-to-psychologists-5180378

Sharma AP (2021). Consumers' Purchase Behaviour and Green Marketing: A Synthesis, Review and Agenda. *International Journal of Consumer Studies* 45(6), 1217–1238

Steiner ND (2016). Comparing Freedom House Democracy Scores to Alternative Indices and Testing for Political Bias: Are US Allies Rated as More Democratic by Freedom House? *Journal of Comparative Policy Analysis: Research and Practice* 18(4), 329–349

Sullivan GM and Artino AR (2013). Analyzing and Interpreting Data from Likert-Type Scales. *Journal of Graduate Medical Education* 5(4), 541–542

Thurstone LL (1927). A Law of Comparative Judgement. *Psychological Review* 34, 273–286

Thurstone LL (1928a). Attitudes Can Be Measured. *The American Journal of Sociology* 33(4), 529–554

Thurstone LL (1928b). The Measurement of Opinion. *The Journal of Abnormal and Social Psychology* 22(4), 415–430

Tidy A (2014). Rich and Poor Alike Are Worried about Income Gap. *CNBC*. Available at: https://www.cnbc.com/2014/09/24/rich-and-poor-alike-are-worried-about-income-gap.html

Transparency International. *Corruption Perception Index Data and Reports*. Available at: www.transparency.org/en/

Ura K, Alkire S, Zangmo T and Wangdi K (2012). *An Extensive Analysis of GNH Index*. Centre for Bhutan Studies, Thimphu, Bhutan

Ursu RM (2018). The Power of Rankings: Quantifying the Effect of Rankings on Online Consumer Search and Purchase Decisions. *Marketing Science* 37(4), 507–684

Vagias WM (2006). *Likert-Type Scale Response Anchors*. Clemson International Institute for Tourism & Research Development, Department of Parks, Recreation and Tourism Management. Clemson University

van Meel-Jansen A (2006). The Magical Number Five in Art Appreciation. *Empirical Studies of the Arts* 24(1), 107–118

Wilhelm PG (2002). International Validation of the Corruption Perceptions Index: Implications for Business Ethics and Entrepreneurship Education. *Journal of Business Ethics* 35(3), 177–189

Willits FK, Theodori GL and Luloff AE (2016). Another Look at Likert Scales. *Journal of Rural Social Sciences* 31(3), 126–139

World Bank (1997). *The State in a Changing World. World Development Report 1997*. Oxford University Press, Oxford

World Bank (2024). *Worldwide Governance Indicators*. Available at: www.worldbank.org/en/publication/worldwide-governance-indicators

World Bank. *World Bank's Living Standards Measurement Study (LSMS)*. Available at: www.worldbank.org/en/programs/lsms

World Justice Report. Rule of Law Index (2023). Available at: worldjusticeproject.org/rule-of-law-index/

Young JL (2017). Numbering the Mind: Questionnaires and the Attitudinal Public. *History of the Human Sciences* 30(4), 32–53

4

PARTICIPATORY APPROACHES TO DEVELOPING SUSTAINABILITY INDICATORS

4.1 Introduction: Why participation?

One commonly applied approach to the development of perception-based indicators is to ask the opinion of experts. Those who develop and are responsible for giving values to the indicators, the indicator "curators", may be based in international agencies, governments, companies or non-governmental organisations, but in each of the examples shown in Table 3.1, the "curators" were either an expert or a group of experts, albeit responding to what they saw as a "need" for the indicator(s) based on some driver they had identified. The goal is for the indicator to help capture complexity and convey that information to those who can make use of it and thus, hopefully, bring about a desired change. The "use" of the indicator may be instrumental in the sense that people use it to help plan and implement decisions in a direct (cause–effect) mode—"*the indicator tells me this so I must do that*"—or maybe the "use" is indirect in the sense that it helps raise awareness of the issue the indicator is designed to reflect—"*the indicator has shown me that this is an important issue so maybe I need to do something about it*". I have argued throughout this book that, despite appearances, all indicators have an element of subjectivity that underpin them. They are, by definition, simplifications of an underlying complexity and assumptions are necessary in terms of choice of indicator, how to select and collect the data that is required for "populating" the indicator and how the calculations need to be made, to name but three. Altering the design or methodology behind the indicator can vary its value and thus change the nature of any guidance it provides.

Many of the perception-based indicators covered in Table 3.1 are, in practice, populated by instruments widely used in the social sciences such as questionnaire-based surveys (e.g., the happiness indicators) or perhaps online-based approaches such as

DOI: 10.4324/9781003456605-4

the Numbeo example. There is obviously much logic here given that these indicators are based, at least in part, on perception, and the best way of getting a sense of that perception is by asking people! In many cases, the questionnaires have what is called a "structured" design with questions which invite respondents to select from a list of answers or perhaps to provide a rank or score based on the Likert scale. These results can then be summarised as counts of respondents answering a certain way or, for answers based on scales and ranks, by the calculation of statistics such as the median and range. Questionnaires can also have a more semi-structured design, with a mix of structured questions along with more open-ended questions designed to draw out why people have respondents the way that they have. There is a spectrum of questionnaire types ranging from structured at one end right through to unstructured, with the latter perhaps comprising nothing more than a set of prompts as a starting point. Structured questionnaires have the advantage of providing data that can be tabulated, graphed and analysed using statistics. The more unstructured type of questionnaire allows us to form a more "go with the flow" type of conversation and deep exploration of what respondents think and why. With the more structured approach to questionnaires the respondent has little, if any, control over the whole process, and they are being asked to respond to questions already set by the researcher; the respondent cannot change those questions or indeed the choice of answers they are presented with. The unstructured approach does provide more freedom for the respondent to take the conversation where they want to. Questionnaire design is something of an art form, and the interested reader is referred to texts such as by Bolton and Brace (2022) for more information.

While the use of the questionnaire-based survey as a research tool has advantages such as allowing for a clear focus on research objectives, there are other approaches to engaging with people that could be taken. For example, a group of people with much experience and knowledge of an issue (i.e., those in the "know") could be invited to explore the issue under investigation, perhaps with minimal guidance from the researcher. This more discursive "group-based" approach is referred to as "participatory", although it is perhaps a rather odd term to use given that questionnaire-based surveys are also "participatory" in the sense that people participate in them. The difference is that with participatory-based approaches the respondents are in the same space at the same time and thus can hear and discuss what each other are saying and respond. With questionnaire-based approaches, even with those at the more unstructured end of the spectrum, the process tends to be more one-to-one. With participatory approaches, the intention is to provide space for people to discuss and express their views and insights, and it is the participatory approach applied to the development of indicators and, in particular, perception-based indicators that is the subject of this chapter.

The chapter begins with an overview of participation, including their history and rationale, before looking at the range of approaches with an especial focus on one of my favourites—the Soft Systems Methodology (SSM). The chapter moves on to explore the use of soft systems to develop indicators of sustainability, and how perception-based indicators can emerge as part of that. The chapter ends with a discussion of some of the disadvantages of using a participatory approach for identifying indicators of sustainability.

4.2 Participatory approaches: A brief history

Participatory approaches are all founded on the same assumption that providing people with the space to analyse an issue or even to identify the relevant issue(s) in the first place ensures that those who are in the "know" can provide more insightful and useful guidance to help with the planning of research, problem solving or indeed interventions. Participation, or as it is sometimes called "stakeholder participation", has a very long history. Indeed, given that human beings are social animals, it can readily be argued that participatory approaches to problem solving are as old as our species. Some of the oldest rock art we have found of human beings from 28,000 to 6,000 BC in Brazil shows them working together to hunt game. Hence, it is almost impossible to identify any single point in history and claim that participatory approaches to problem solving began there. But another way of looking at this is to consider the history of documented approaches of participation as a formal process designed to help understand complexity and to solve problems. Simon Goldhill of the University of Cambridge has made a case that the first documented account of participatory analysis as a process we would know today is set out in the book "Anabasis" (or "The Expedition of Cyrus") written by the Greek historian and soldier Xenophon (~400 BC). The book sets out the story of a Greek army (10,000 strong soldiers with probably the same number again as camp "followers") trapped deep behind enemy lines in Persia, and which had lost many of its leaders including "Cyrus the Younger" (the overall commander of the army) in battle. The remaining soldiers are left to fend for themselves and find their way back to friendly territory. It is, in essence, a story of survival, as people are forced to learn and adapt, and throughout the story runs a thread of a participatory approach to decision–making:

> True soldiers in their readiness to meet danger and hardship, they were still free Greeks, who could be governed with their own consent and who insisted upon being heard and consulted in matters of common interest. They have been truly called, "a marching democracy", "a roving commonwealth", "deliberating and acting, fighting and voting; an epitome of Athens set adrift in the centre of Asia".
> *(page xiii)*

Xenophon was one of those living within this group, and Anabasis is presented as a first-person narrative and analysis of the social dynamics at play. In the book, he paints a picture of this slow march back home and how people interact amongst themselves as a group but also with local communities with which they must engage with for food and other resources. They observe and learn from these communities, and Xenophon was able to analyse these interactions from the "inside out". Hence, as Goldhill notes:

> I think it's interesting to compare back to Herodotus the great history of the Persian Wars, who, as it were, founds our sense of the Orient of the East as this mysterious and difficult place. Xenophon gives us a very different image and a very important and instructive image of how he actually engages with these people. He actually talks to them as they have to buy food. It's the first example of as were participant observation.
>
> *(Goldhill, 2011)*

Anabasis has, incidentally, provided inspiration for a novel called "The Warrior's" written by Sol Yurick and set in a dystopian vision of New York in the 1960s rather than Persia of 400 BC. In the book, a gang called the "Warrior's" finds itself outside its territory and has to fight its way back to its home "turf". Indeed, the book in turn became an inspiration for a 1979 movie called "The Warriors" (poster line "These are the armies of the night"), which is still screened on a regular basis today and often ranks high amongst listings of "cult" movies. It may be a bit of a stretch, but nonetheless, this cult movie classic provides a link back to the very origins of participatory problem solving!

Moving to much more recent times than Anabasis, there are many strands to the development of participatory approaches for generating insights and solving problems that have intertwined and fed off each other during the latter half of the 20th century. It is simply not possible to go through all of them here, and I can only hope to provide a few of the highlights. A summary of key elements within a timeline stretching from the 1940s to this day is provided in Figure 4.1.

The context for the top-half of Figure 4.1 is development in a period of de-colonisation that occurred following the Second World War. The 1950s and especially the 1960s witnessed the birth of many independent states that had formally been colonies of European powers such as Britain, France, Belgium and Portugal. It is as well to remember that this same period also saw an increase in global tensions caused by the "Cold War" between the USA and other western democracies and the Soviet Union and its satellites. The post-war period also saw much optimism about the contribution that technology could make to the development of humankind, and the inaugural speech given by US President Truman in 1949 is often regarded as the starting point for the rise of what we now call "modernisation", which broadly equates to what many regard as development:

> [W]e must embark on a bold new program for making the benefits of our scientific advances and industrial progress available for the improvement and growth of underdeveloped areas.
>
> I believe that we should make available to peace-loving peoples the benefits of our store of technical knowledge in order to help them realize their

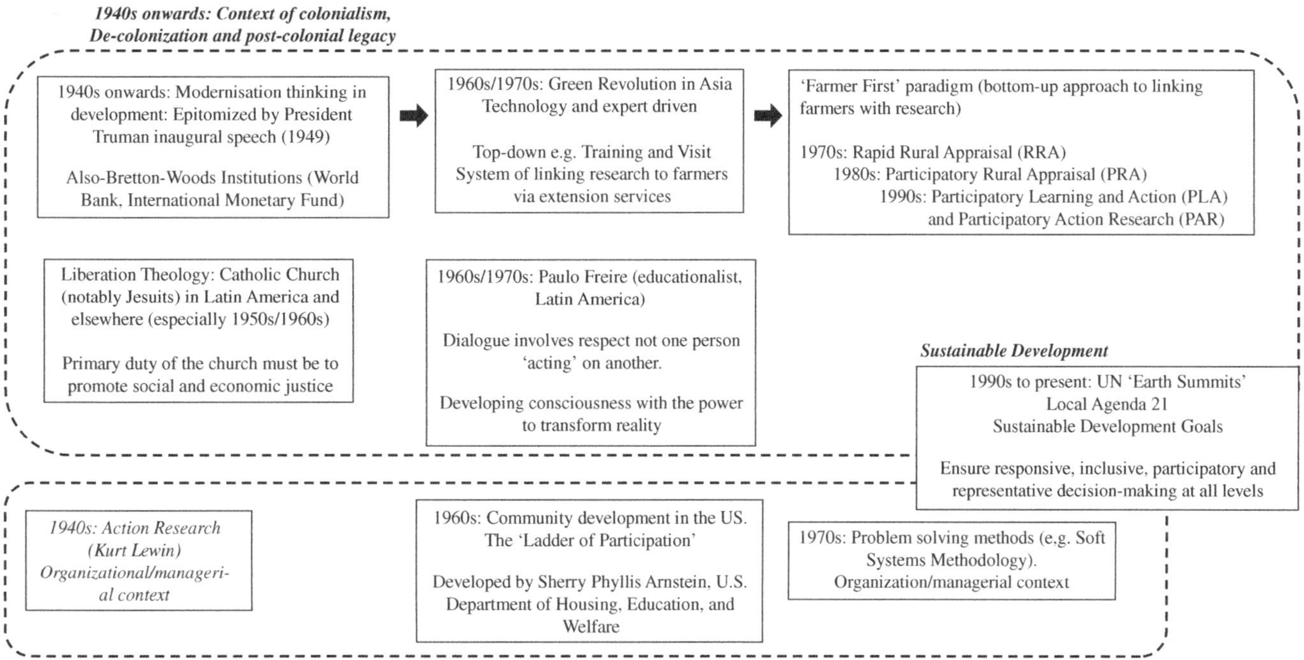

FIGURE 4.1 Some key elements in the rise of participatory approaches for helping to design, plan and implement interventions in a range of sectors

aspirations for a better life. And, in cooperation with other nations, we should foster capital investment in areas needing development.

Our aim should be to help the free peoples of the world, through their own efforts, to produce more food, more clothing, more materials for housing, and more mechanical power to lighten their burdens.

(Truman, 1949)

There was a desire to improve the lives of the people in former colonies, and modernisation was seen as the way to achieve this. Perhaps, the most obvious expression of that drive for modernisation was in the "Green Revolution", which took place primarily in Asia, during the 1950s and 1960s (Jain, 2018). It must not be confused with the "green revolution" we often speak of today, which is focussed on a rise in awareness of the importance of environmental issues and adoption of technologies such as renewable energy. Instead, the Green Revolution from the 1950s and 1960s was centred on improving agricultural productivity, especially for small-scale farmers, and its champion is often regarded as being the scientist Norman Borlaug (Hesser, 2006). The programme encouraged the development and supply of new varieties of crops (high yielding, disease resistant, responsive to fertiliser) as well as the promotion of artificial fertiliser and pesticide use by farmers and thus was not designed, at least primarily, to be "environment friendly". In parallel, the programme enhanced agricultural research and "extension" services (i.e., services designed to engage with farmers) to help provide information and feedback. But these "solutions" to boosting crop yields were those identified by agricultural experts and scientists and were designed to address what they saw as the key challenge to productivity faced by farmers in the developing world which resulted in poor levels of food security. Hence, the interventions that were put into place were essentially "top down" even if, in theory, there was an accommodation for a two-way flow of ideas and insights between researchers and farmers—as exemplified by the "Training and Visit" system (T&V) of extension that was widely used in the "Green Revolution" (Benor and Harrison, 1978).

The "Green Revolution" had not only its successes but also its failures. An example with the latter was a widening gap between the richest and poorest in society. Those farmers having the most resource and often the best education/training were able to make better use of the inputs and advice compared to those with less in the way of resource and education/training. Added to this, of course, were environment issues associated with an increase in fertiliser and pesticide use. There was also much frustration with the Green Revolution being seen purely through an economic lens so that success was measured via contributions to national GDP rather than how it helped improve the livelihoods of the poor. It was not long before development practitioners and researchers began to question the "top down" nature of the "Green Revolution" and argued for the voices of farmers articulating their needs to be heard (Shiva, 1991). The response in the 1970s was a call for more participation from those intended to benefit from the Green Revolution,

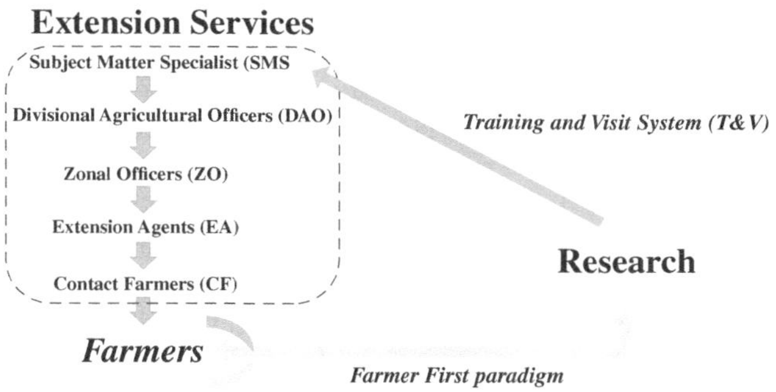

FIGURE 4.2 The shift from a top-down interaction (the T&V) with farmers to a participatory "farmer first" approach (adapted from Benor and Harrison, 1978)

and this resulted in an evolution of approaches and tools that came to be grouped under the clarion-call of "Farmer First"" (Chambers, 1983, 1994, 1997, 2017). The emphasis on farmers being first is not surprising given that most of the poor in the developing world were living in rural areas and had livelihoods based on farming. Indeed, this awareness was at the heart of the "Green Revolution". The shift in emphasis from "top down" design of interventions to a "farmer first" approach is summarised as Figure 4.2. This figure presents two ways in which farmers can connect with research. The first of these, and indeed the version that had been widely adopted in the Green Revolution, was based on the "T&V" model of extension (Benor and Harrison, 1978). Here, the results of research are interpreted by "Subject Matter Specialists" who then tease out what they regard as the key information (i.e., the messages) that need to be passed to farmers through a chain of extension agents operating at different geographical scales within the community. In theory, the arrows running from researchers to the farmers should be bi-directional, so that researchers can address the concerns of farmers, but in practice, the process tends to move in just one direction—from researcher to farmer (i.e., "top down"). As a result, researchers tend to work on what they regard as important, not necessarily what farmers thought was important. In the alternative "farmer first" model, the farmers work directly with researchers, as equals, to co-generate possible interventions that will address the needs of the former. Various approaches were developed to help with the elicitation of information from farmers and their families in terms of what they need and how those needs can best be met, and some (Rapid Rural Appraisal, Participatory Rural Appraisal, etc.) are included towards the right-hand side of Figure 4.1.

The 1960s also saw changes in the traditional model of education which paralleled the "farmer first" story above. Paulo Freire, a Brazilian educationalist (1921–1997), argued for a move away from a "top-down" or what he referred to as the "banking" approach to education with teachers telling students what they thought they needed to know and thus filling them up with knowledge in much the same way that banks fill-up with money (Freire, 1976, 2000). He made the case that education should be about dialogue and dialogue involves respect; not one person (the teacher) "filling" the other (the "pupil") with knowledge, the teacher deems to be necessary. For Freire, a more participatory education model (or problem-posing education) was all about developing consciousness with the power to transform reality and that had to be achieved via a process of teacher and student working together (Schugurensky, 1998).

Another, albeit older, strand that needs to be mentioned is "Action research", which is often said to have originated from the work of Kurt Lewin in the 1940s (Lewin, 1946, 1958; Adelman, 1993):

> Action research must include the active participation by those who have to carry out the work in the exploration of problems that they identify and anticipate. After investigation of these problems the group makes decisions, monitoring and keeping note of the consequences. Regular reviews of progress follow. The group would decide on when a particular plan or strategy had been exhausted and fulfilled, come to nothing, and would bring to these discussions newly perceived problems
>
> *(Adelman, 1993; page 9)*

However, unlike the examples for "Farmer First" and education given above, the focus of "action research" was more upon participation in the context of organisations, including companies, rather than communities in the developing world, but the underlying paradigm is the same. Indeed, Lewin argued that research needs to be seen more as a basis for social action designed to help people:

> The research needed for social practice can best be characterized as research for social management or social engineering. It is a type of action-research, a comparative research on the conditions and effects of various forms of social action, and research leading to social action. Research that produces nothing but books will not suffice.
>
> *(Lewin, 1948; page 202)*

This emphasis of "action research" on learning within organisations is one we will return to later in the chapter.

There are many other strands to the rise of participation for problem posing and solving, far too many to be covered here, but it does need to be stressed that community participation is very much at the heart of sustainable development. The first "Earth Conference" facilitated by the United Nations and held in Rio de Janeiro in 1992 had what was called "Local Agenda 21" at its heart with the strapline of "Think global. Act local". LA21 embraced community participation, and indeed, there were many programmes throughout the globe where local communities were encouraged to define how they wanted to contribute towards sustainability, and as part of this, they developed sets of SIs to help conceptualise what they needed, as well as tracking progress towards goal they had set (Bell and Morse, 2008). Indeed, the centrality of community participation can be seen to this day in the SDGs even if there is no single SDG having “participation” in its title, although there is mention of the need to develop partnerships (SDG 17) and inclusivity (SDG 16) in two of them. Participation is at the heart of the SDGs.

4.3 The landscape of participatory approaches: From focus groups to soft systems

In line with the gradual rise in importance of participatory approaches in the 1960s, the end of that decade witnessed the publication of a framework to help illustrate differences between types of participation. Sherry Phyllis Arnstein of the U.S. Department of Housing, Education, and Welfare developed what she called the “Ladder of Participation”, and a summary is provided in Figure 4.3 (Arnstein, 1969). Figure 4.3a is the ladder, with types of participation set out as the rungs. Figure 4.3b is a diagrammatic representation of the Arnstein types of participation based on a pictorial device created by the International Association of Public Participation (IAP2).

The concept of a “ladder” of participation has been highly influential, and the fundamental point Arnstein was making is that participation is not solely about the methods used to engage people, but why it is being done, the form of relationship between those involved and how the outcomes of the process are used, if at all. At one extreme, towards the bottom end of Arnstein’s Ladder, participation is nothing more than manipulation, perhaps even exploitation. At this extreme, the person running the process is only seeking to engage with people to manipulate them to agree to a preconceived set of ideas. The participants have no power whatsoever in this scenario; all the power rests with the person running the show. Indeed, not only that but also it feels nefarious; the engagement of people is being used to promote an agenda, whether they like it or not.

Above “manipulation”, the next rung of the Arnstein ladder is “informing” whereby decisions are made by others, and the local community is simply informed of them. Some may call this “outreach” or perhaps even “communication”, but it is in essence a one-way process even if the community is allowed to ask questions or seek clarification or some points. The decisions have been made so again that there is very little power resting with the community although at least here there is, hopefully, no deception involved and those informing the community of decisions are being genuinely open and transparent about what they are and the impacts that may arise. The T&V process of agricultural extension noted above and in Figure 4.2 is an example of participation implemented as “informing”. There may be some fine lines here between “manipulation” and “informing”, and much may depend on the perspective of the “givers” and “receivers” of the information. What some

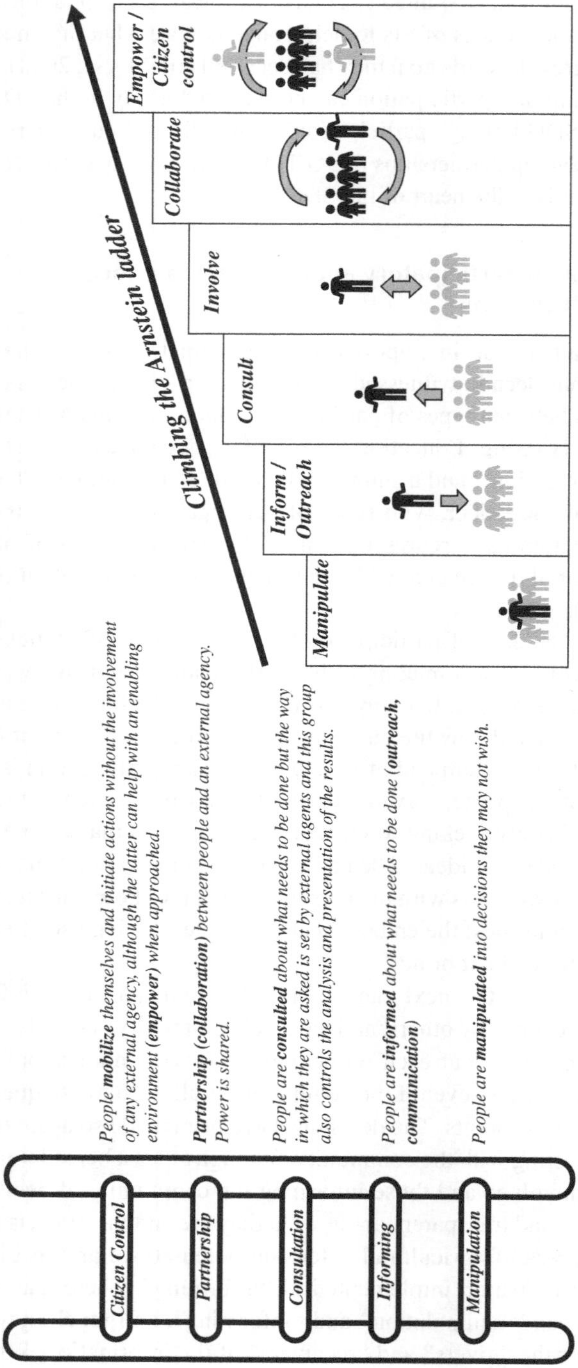

(a) The Arnstein Ladder

(b) Types of participation. Based on the International Association for Public Participation (IAP2) framework.

FIGURE 4.3 Arnstein's Ladder of Participation and a diagrammatic representation of the types of participation (after Arnstein, 1969)

may view as well-intentioned "informing" may be seen by others as an attempt at "manipulating" them into doing something they may not wish to do.

The next upwards "rung" of the ladder is called "consultation", and here the community is asked for its views and opinions before final decisions are made with the expectation that these will influence what is being planned. The process of "asking" is still controlled by external agents as indeed is the decision-making. Hence, there may be no guarantees that the views of the community will be listened to, and the process may be more of a "tick box" exercise where those in charge can at least say that they "consulted" with the community even if they did not listen to what they had so say. As above, there may be fine-lines here and if the consultation is done in name only, with no intention at all to make use of the feedback, then the process looks more like "informing", maybe even tinged with exploitation. There can also, of course, be more structural issues at play, which prevent feedback from consultation being effective. In theory, the T&V process was founded on consultation as the views of farmers were meant to be fed back to the researchers and help guide their activities, and the intention was genuine, but in practice, it was often a one-way process of "informing".

The next rung on the ladder is called "partnership" (or collaboration), and here the community will work together with external agents to help identify their issues and bring about desired changes. The assumption is that external agents do not come with a priori decisions, and instead, they consult with the community to see what they think. The agents work with the community from the start to help identify desired outcomes and what needs to be done and how best to achieve them. Power is shared between the partners, external agents and community, and they work together.

At the very top of Arnstein's Ladder are approaches to participation initiated and run by communities themselves without any, or perhaps minimal, support from outside experts. In this case, the participation is internal to the community, and they own it as well as organising themselves to help put the insights that have been gained into practice. They may, or may not, invite someone from outside to help them arrive at their insights or perhaps provide information when asked but the key point here is that the community is in charge.

The Arnstein's Ladder has been a popular device and is a useful way of reminding us that "participation" comes in many flavours, and while the term is often bandied around, it can mean many different things to different people. Perception is infused throughout Figure 4.3, and even towards the top end of the ladder, there is still scope for manipulation and bias as some may perceive and believe that they are in a "partnership", while in practice, they are being manipulated in clever ways that may not be readily apparent to all involved. The rungs in Arnstein's Ladder are perhaps not as solid and well-defined, as we may think or perhaps wish them to be, and there can be much nuance that means the rungs are far "fuzzier" than implied by the sharply drawn lines in Figure 4.3. It also needs to be stressed that the ladder is not intended to be an objective scale of superiority running from manipulation as the least desired to empowerment as the "best", although it is often interpreted that way. "Citizen control" and "partnership" may not necessarily always be "better" or more desirable than "informing". It does depend on context, and there are times when decisions must be made for a "greater good" that go beyond the wishes of a local community. Hence, there are times when "informing" can be the better and indeed more honest option. But even so, having "manipulation" at the bottom of the ladder and thus the most undesirable approach to participation does feel right. Since Arnstein first published the "ladder" model, there have been various attempts to build upon it, and, for example, Geekiyanage et al. (2021) provide a classification of methods often used for participation and how they align with the Arnstein categories (Figure 4.4).

There are various methods and tools that can be used to "inform" communities, including the use of websites, printed material (leaflets, posters) and exhibits. For "consultation", there are tools such as interviews, surveys and crowdsourcing, all of which are mentioned in Chapter 3, but also focus groups. Focus groups are similar in style to interviews, but in this case, a group of respondents are present in the room at the same time, and as a result, this approach—a sort of "group interview"—can be time and cost-effective compared to individual interviews. The facilitator of the focus group controls the process as they set the questions to ask and manage the discussion. As can be imagined, the size of a focus group is important; if it is too big, it can become unwieldy, and a maximum size of ten people is often cited. The advantage of the focus group is that participants can listen to the

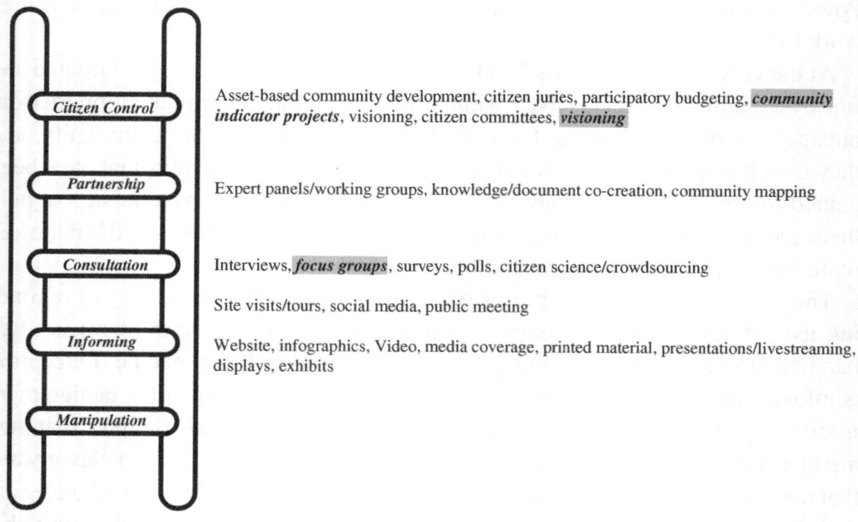

FIGURE 4.4 Arnstein's Ladder of Participation and the classification of participatory approaches

Adapted from Arnstein (1969) and Geekiyanage et al. (2021).

views of others and that may help frame their own thinking. It is also possible for the facilitator of the focus group to assess the body language of those taking part. Is there unease, for example, amongst some participants when comments are made by others? One of the disadvantages of focus groups is that the group can be captured or steered by a few dominant individuals, and as a result, there can be what is called "Groupthink" where everyone becomes influenced to go along with a particular view even if some have concerns. Another potential issue with focus groups is bias from the moderator who could either consciously or unconsciously steer (manipulate) the discussion in a particular way to arrive at a desired result. In fairness, this is potentially an issue with interviews or indeed any participatory process where the facilitator takes the role of directing the discussion.

The "Partnership" rung of the ladder may involve the use of expert panels and working groups as well as community mapping. Finally, at the top rung of the Arnstein's Ladder, there can be tools such as citizen committees and citizen juries. Also included here are tools that involve "visioning" by communities, and an example is provided by the SSM (Checkland, 1998, 2000; Checkland and Scholes, 1999; Checkland and Poulter, 2006) covered later in the chapter in greater depth. At this top end of the ladder, an external agent may act as a facilitator of the process but play a relatively minor role—more as a guide or facilitator to help participants. In the SSM, for example, the facilitator does little more than present the steps of the process to the participants and does not engage in the deliberations. This more "hands off" approach does not eliminate the potential for bias on the part of the facilitator, but it does help to limit it. Also, it does need to be said that while Figure 4.4 sets out some of the tools that can be used for the types of participation, this is a rather simplistic classification, and in practice, a single approach may combine them. For example, a citizen jury (classified under "citizen control" in Figure 4.4) may well involve site visits and tours, and a partnership between a community and external actors may decide that they need to collect new data via a survey and either implement it themselves or hire a specialist.

All the participatory approaches shown in Figure 4.4 can be used to develop SIs (Bell and Morse, 2003, 2008, 2014; Morse, 2015). In addition, indicators can be used to help inform communities and indeed can be a useful means by which sustainability concepts are framed and understood (Morse, 2008). Sustainability can be a rather nebulous concept, so pinning it down to some key elements and associating them with indicators can help provide clarification and help with the monitoring of what the community regards as key issues that need to be addressed. However, the Arnstein's Ladder, and the issues associated with it, still apply. A community may be manipulated into agreeing to indicators set by experts, or they may simply be communicated to the community with them having no real choice other than to accept them. Further up the Arnstein's Ladder, indicators may be developed in a more collaborative or partnership mode, and at the very top of the ladder, the community may set its own indicators with or without any advice from experts. Perception-based indicators could well be a part of the mix here, and it is possible that a community may come up with such indicators without necessarily thinking through what data may be needed to populate them and how they would get these data.

The remaining sections of the chapter will focus more specifically on SSM as an approach to participation and how it can help formulate SIs. I will not go into great depth regarding SSM, and its application in sustainability as fuller descriptions and discussions can be found in other publications (Bell and Morse, 2003, 2008, 2012). Instead, it is used here as an illustration of how it has been used for identifying SIs and the issues that may be involved. The examples will draw from projects I have worked on, along with the use of SSM to engage with postgraduate students at the University of Surrey, UK.

4.4 Introduction to the SSM for developing SIs

The first point that needs to be made is that SSM is a flexible approach to participation and can be used, as with all the approaches set out in Figure 4.4, to address a wide variety of issues and contexts (Bell and Morse, 2003, 2008, 2012). It is by no means an approach that is focussed solely on indicators or even, for that matter, sustainability. Its origins were very much in the world of organisations, just like action research referred to above, and were designed to help provide a way in which staff working within them could be harnessed to get a better understanding of the issues they face and how best to overcome them. The premise was a simple one. When it came to improving the functioning of an organisation to help it deliver products and services (healthcare, education, transport, etc.), the issues that needed to be addressed could be complex and may not necessarily boil down to a quick technical fix such as a new piece of software or machinery. The issues involved may be highly complex and resist simplification (i.e., the issue is "wicked"), and people's perceptions of this complexity may also vary. The term "wicked" in this context is not meant to imply that these challenges are evil; just that they are complex and challenging to dissect and understand. Clearly, there is need to identify and unpack the issues before changes can be planned and implemented; otherwise, the changes may have no impact or may even make matters worse (Checkland, 1998, 2000; Checkland and Scholes, 1999; Checkland and Poulter, 2006).

SSM comprises a series of steps, and the participants need to be guided through the steps in order for them to make the most of it, and this is the role of the facilitator. With an external facilitator, there is clearly a challenge as they need to keep out of the unpacking and learning as much as possible while at the same time acting as a trusted and independent guide. It is entirely possible for the group to self-steer its way through the SSM process following a sort of "cookbook" approach. Nonetheless, whether SSM is facilitated or self-steered, the process of unpacking (understanding) and coming up with potential interventions must be left to the participants. It is this "control by the group for the group" that rests at the heart of SSM and places it towards the top-end of Arnstein's Ladder.

Here, I will focus on what can best be described as a simplified or "cut down" approach to SSM and especially how it can be used to identify indicators to assess sustainability and whether perception-based indicators can emerge from this (Bell and Morse, 2003, 2008, 2012). I will illustrate the SSM steps using "real" outputs from a series of workshops with postgraduate students at the University of Surrey, all of whom were taking postgraduate degrees in sustainable development. The students were invited to analyse sustainability at the university and what they think could be done to improve matters. The University of Surrey is a campus university located on the outskirts of Guildford, a medium-sized town in the county of Surrey, South-East England. The students were divided into groups, labelled for convenience A, B and C, of typically five–seven members. However, while they had much in common in terms of their interest in sustainability, their chosen degree programme, age (most were in the early 20s) and country of origin (most were from the UK), they did differ in terms of their first-degree subject. An outline of the SSM process the student groups followed is shown in Figure 4.5, and the outputs provided below for illustration are from workshops held between 2012–2013 and 2015–2016.

There are four steps in Figure 4.5 divided into three phases. The first phase can be called "what is?" and involves the group arriving at an agreed, amongst themselves, understanding as to the current state-of-play of sustainability at the university. Almost inevitably, this will result in an analysis that spans many actors, issues and processes. Each group are encouraged to create a RP which captures all that they regard as important and relevant. The RP is, as the name suggests, a pictorial representation of their analysis, using as little text as possible, and some examples can be found in Bell et al. (2016a). Examples from the student SSM workshops are

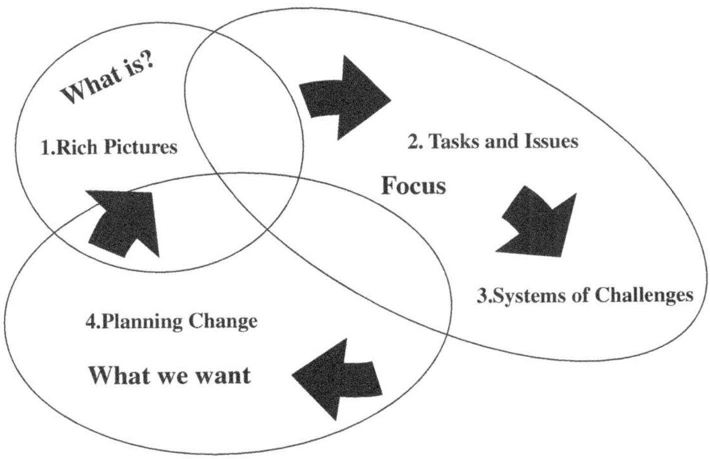

FIGURE 4.5 Summary of the steps involved in a soft systems-based analysis (after Bell and Morse, 2008, 2012)

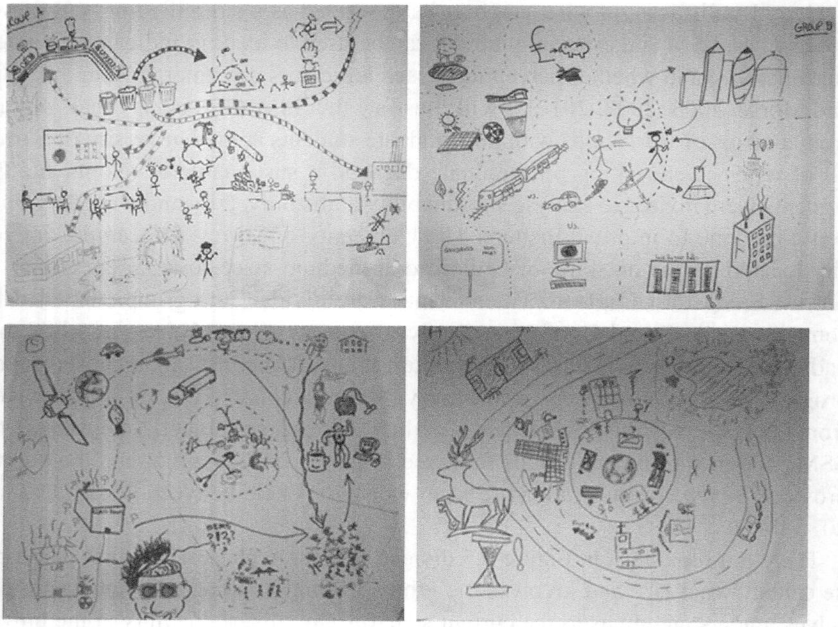

FIGURE 4.6 Examples of RPs developed by student groups asked to analyse sustainability at the University of Surrey

provided in Figure 4.6. Without wishing to go into the details of each RP, there is clearly much in them, and even a casual glance at the pictures suggests a spanning of the global and local, as well as many topics such as education, finance, energy, food, housing, transport, biodiversity and social issues such as security and sports. Such complexity is what we would expect to see in such analyses; sustainability, after all, covers just about all aspects of human life, and its interaction with the world around us and even the analysis of just one institution would generate many elements.

But there is also diversity. For example, there are concerns expressed in some of the RPs over personal safety and issues of "Fairtrade" surrounding the availability of food and drink on campus, but this does not appear in others. Some groups take the analysis of sustainability to the global level and include the recruitment of overseas students and how that has an impact on the "financial sustainability" of the university, as well as the cultural diversity of the student body, while others tend to have a stronger focus on the UK policy context. There is no "right" or "wrong" answer here; the RPs are the analyses based on the experiences and perceptions of students within the groups, and diversity is inevitable. Indeed, as the exercise is group-based, the RPs in Figure 4.6 are a negotiated outcome of sharing perspectives.

In the second phase, called "Focus", the students are requested to identify what they see as the key issues emerging from their analysis along with tasks that they feel need to be undertaken. Thinking in terms of issues and tasks allows participants to prioritise in ways they feel most comfortable. For some participants, it may be more convenient to think in terms of key issues that need to be addressed, while for others, it may be easier to express priorities in terms of tasks. These are related of course; an issue may well have a task, or perhaps a couple of tasks, associated with it. For example, a group may identify what they regard as a lack of public transport to and from the campus as an issue. Another group may have the same view but express it as a task—"improve the availability of public transport". One straightforward way of identifying key issues and tasks is to write them on "Post It" stickers (maybe with different colours for tasks and issues) and place them on the RP where they occur.

Issues and tasks are likely to differ in terms of their importance and indeed how challenging they will be to address in practice. Hence, once the key tasks and issues have been identified, each group is encouraged to place them in a two-dimensional matrix as shown with the steps in Figure 4.7. The horizontal axis represents relative importance of the task/issued, while the vertical axis is the relative ease of implementation. The top right-hand side of the matrix represents those tasks/issues that are deemed by the group to be most important and easier to put into practice. At the bottom left-hand corner of the matrix, the other end of the scales are those tasks/issues the group regards as being of lesser importance and more challenging to implement.

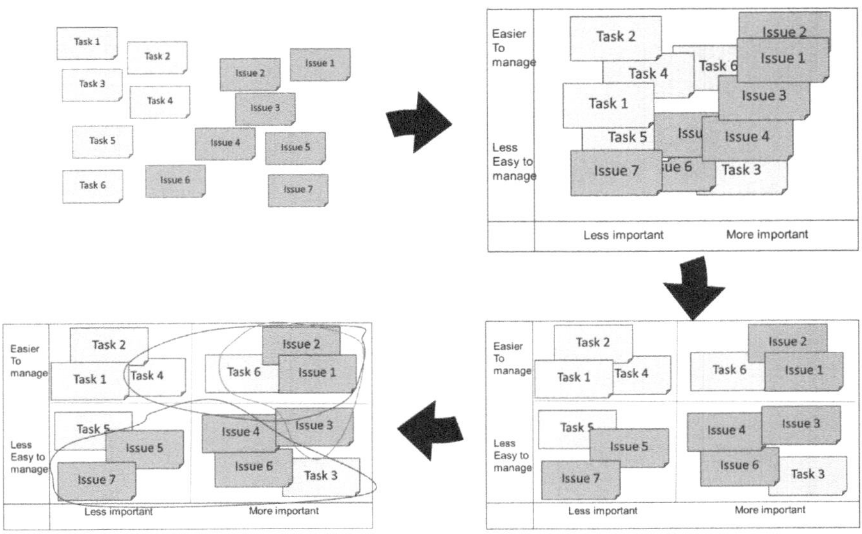

FIGURE 4.7 Identifying key tasks and issues from RPs (after Bell and Morse, 2008, 2012).

In the final step, lines are used to cluster related tasks and issues identified by the groups. These clusters are given names and for the basis for planning interventions.

Some examples of the final matrix of tasks and issues for the students analysing sustainability at the University of Surrey are shown in Figure 4.8. It is these tasks and issues that form the basis for interventions which the students feel need to happen to bring about their desired change at the university.

It is possible for different issues/tasks to be related. For example, sticking with the transport theme, a group may wish to see more cycle lanes along roads near to the campus, and this could overlap with another issue such as the need to engage more with local communities as their agreement could be critical for success. These related tasks and issues may well cut across the four segments of the matrix, and the next step involves inviting each group to draw lines around tasks and issues they feel are related and give that cluster a name. Related tasks and issues in these clusters can be thought of as the basis for desired interventions, or what can be called "proto-interventions" or "proto-projects", and the groups can use the clusters to begin to map out what needs to be done, for what purpose and by whom. This is, of course, just a first step, and there are other dimensions that need to be considered before these "proto-interventions" can be implemented, not least being any resources that may be required, but at least, they represent ideas to help bring about a more sustainable university.

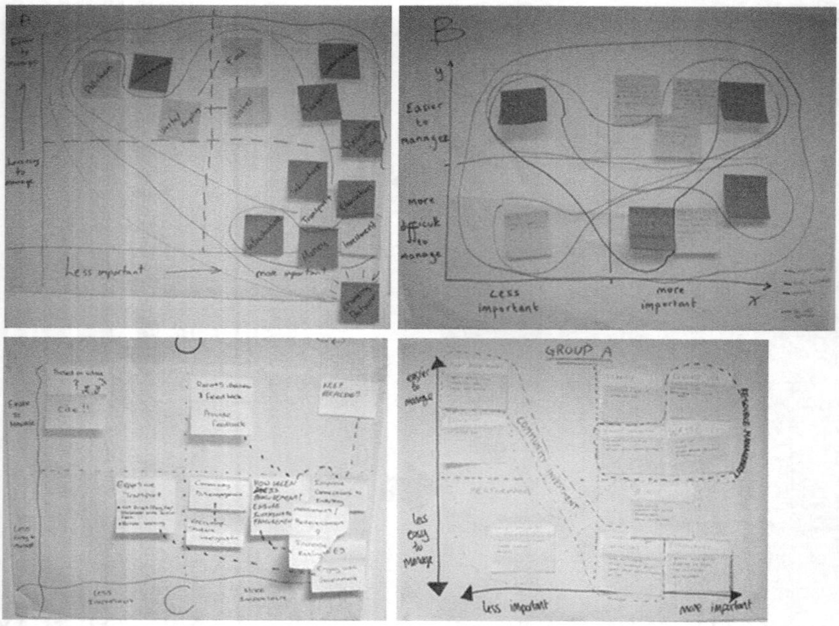

FIGURE 4.8 Examples of Systems of Challenges derived from key tasks and issues identified by student groups asked to analyse sustainability at the University of Surrey

To help progress each of the proto-interventions set out in Figure 4.8 to something that is implementable, it is important to address a set of consideration that can be expressed using the BITOAC (Beneficiary, Implementors, Transformation, Owners, Assumptions and Constraints) mnemonic checklist as illustrated in Table 4.1. The BITOAC checklist is one of many such devices, and another popular one is CATWOE (Customer, Actor, Transformation, Worldview, Owner, and Environment). Each group is invited to come up with answers to the checklist for each of their proto-interventions, and it is possible, indeed likely, that even for the same "proto-intervention" views may vary. For example, there may well be a range of views even for the fundamental question as to who is intended to benefit from an intervention. The BITOAC checklist can also provide a basis for the development of indicators, especially when it comes to the Transformation, Assumptions and Constraints components, and expressing these as indicators can help facilitate discussion within the group. At this stage, the indicators may act more as tools for framing discussion around the elements of BITOAC and thus do not need to be developed in terms of definition, methodology, etc.

TABLE 4.1 The BITOAC framework for planning interventions (after Bell and Morse, 2012)

Letter	*Term*	*Questions*
B	Beneficiary	Who is the Beneficiary? Who is meant to gain from the intervention? These will often have a high level of interest in the intervention but not necessarily have all that much power to influence them.
I	Implementors	Who are the Implementers? Who will be responsible for doing the work in the intervention?
T	Transformation	What is the Transformation? What is the nature of the intervention? What is meant to be done and what will need to be achieved to help the beneficiaries?
O	Owners	For example, this is often the person or organisation who provides the resources for the intervention or perhaps someone acting on their behalf. They will be expected to have a lot of power when it comes to making the transformation a reality.
A	Assumptions	What are the Assumptions behind the transformation? What are the assumptions that underpin what is being planned? For example, statements along the lines of "if we do this . . . . Then it is assumed that the following will happen . . ." or "for the transformation to be effective then it is assumed that the following will be provided in a timely manner . . ."
C	Constraints	What are the factors that could potentially hinder the transformation and stop it from having its desired outcomes? An often-cited constraint in this context would be the availability of resources (finance, staff, materials, etc.)

4.5 Participatory approaches to indicator development: Some examples

In Chapter 2, it was shown that only a relatively small proportion (4%) of the 231 SDG indicators are based on assessing perception. But in this case, the process of selecting indicators is driven by experts, namely the IAEG-SDGs. However, participatory approaches to problem solving, such as SSM, can be adapted in many ways, and one of these is their use to help identify and populate SIs (Bell and Morse, 2003, 2008, 2014; Morse, 2015; Talubo et al., 2022). Could more participatory processes, with a wider range of people many of whom may not necessarily be indicator experts, lead to greater emphasis on perception-based indicators for sustainability? It is entirely possible that at least some of the indicators emerging out of a participatory process could be based on a need to gauge people's perceptions, but at the same time, much may depend on the rung of the Arnstein's Ladder where the participatory process is located. Towards the bottom end of the ladder where much control is with external experts, it seems more likely that they will tend to opt for familiar and well-established indicators, perhaps those in the "reality" category described in Chapter 2. Towards the top end of the Arnstein's Ladder where more control and power are in the hands of participants and much less in the hands of experts, it could be more likely that perception-based indicators will emerge as people think "outside the box". At the top end of the ladder, participants may perhaps be more likely to think first about what indicators are needed, and if that requires indicators based on perception, then fine. Therefore, it may be hypothesised that participatory approaches towards the top end of the Arnstein's Ladder, where participants are given more power and flexibility, are more likely to generate perception-based indicators.

The hypothesis set out above has yet to be tested, but perception-based indicators have emerged out of the SSM process, and an example can be found for a project designed to develop indicators of sustainable development for Malta (Bell and Morse, 2003). The project took place in the early 2000s around the time that Malta joined the EU, and a series of SSM workshops were undertaken over the period of a year or so, with each workshop comprising groups working on themes deemed by the Maltese government to be important for sustainable development. The government felt it needed indicators of sustainability for each of these themes, especially given the accession of the country to the EU. The themes they opted for were:

1. Sustainable coastal management
2. Marine conservation areas
3. Integrated water resource management
4. Erosion/desertification control management
5. Tourism: Impacts on health

These themes are not especially surprising given that Malta is an island state, with a relatively small land area and long coastline with fishing and tourism as major industries. Theme 3 is linked to the country's reliance on groundwater from aquifers; although at the time of the SSM workshops, there was one desalination plant (Malta currently has four plants with a fifth under construction). Theme 4 relates to climate change, an important issue for a small-island state, and the abandonment of agricultural land leading to increased erosion and desertification.

The groups were asked to come up with a list of SIs for their allocated theme by following much the same process as set out in Section 4.3 for the students exploring sustainability at the University of Surrey. As befits the modus operandi of the SSM approach, the groups were in charge and free to come up with whatever indicators they thought to be most appropriate for their theme. During the workshop, all groups were asked at various stages to present and explain their assumptions and decisions.

It is not necessary to go through all the indicators that emerged from the workshops, and the reader can find them in Bell and Morse (2003), but the list developed by the team who focussed on the tourism and health theme is shown ins Table 4.2. Ten indicators are listed in Table 4.2, and some of them are to be expected such as the incident of gastroenteritis (cases and outbreaks) and faecal coliform count of bathing water. Gastroenteritis cases can be readily obtained via medical records (an outbreak is called when the number of cases exceeds a certain threshold), and the presence of faecal coliforms can be ascertained via routine sampling of seawater. There are also two indicators in the list focussed on rodents, one of which is a proxy measure of rodent population (number of captures from traps that are set), while the other indicator attempts to capture the level of complaints about rodents from tourists. Given the obvious health risk that rodents provide, the inclusion of an indicator designed to be a proxy measure of their

TABLE 4.2 List of indicators for health and tourism in Malta (after Bell and Morse, 2003)

Indicator	*Note*
Gastroenteritis outbreaks in North-West of Malta	Number of total outbreaks in a year
Gastroenteritis cases in North-West of Malta	Number of total cases in a year
Bathing water quality	% of samples meeting accepTable levels of faecal coliforms (< 1000 mg/l)
Rodent control—bait placed	Number of takes from baits places
Rodent Control—residents' complaints	Number of complaints from residents
Risk factor grading of hotels	% of inspections falling within Grades A–C
Risk factor grading of catering establishments	% of inspections falling within Grades A–C
Beach quality as compared to Blue Flag	In % points attained
Beach quality as compared to health criteria	% of total beach sample observations falling within Grade A
Media interest in tourism health	% of positive/negative media coverage

population is understandable, but why is there an indicator based on complaints? An important consideration which emerged within the group is the sensitivity that people have towards seeing rodents. Even if the population of rodents may be low, such encounters, even if relatively rare, are perceived by the tourists to be serious, and this can have repercussions for the reputation of tourism on the island. Hence, the group decided that a proxy indicator for assessing the rodent population is not enough, and it was important to try and capture complaints. The list in Table 4.2 also includes four indicators based on established frameworks to assess risk factors for hotels and restaurants (e.g., their food storage and handling procedures, cleanliness, level of maintenance) and quality of beaches. These are all established indicators already used by agencies in Malta and are based on regular inspections of beaches and premises and employ a grading system. The "Blue Flag" system of grading beaches is a well-established international system of criteria, first introduced in 1985, spanning factors such as water quality (www.blueflag.global) and has some overlap with other indicators in the list. It was selected for inclusion because of its international status, but the team felt that the other more local indicators, such as the one linked to health criteria, also needed to be included.

The final indicator in Table 4.2 is, in many ways, the most intriguing. The team concluded that media reporting of tourism and health in Malta was important for their theme, and this should be based on assessing the balance of "positive" and "negative" stories. However, this raises some important methodological considerations. First, what media should be considered (the group emphasised newspaper reporting but what about broadcast media?), how many media reports need to be included in the sample and whether the focus should only be on media originating within Malta? The latter is especially important given that the tourists are coming from many countries, and their choice of Malta as a destination could well be influenced by media reports within their own countries, but taking such an international approach would complicate any assessment and make it far more expensive. Social media platforms (e.g., bulletin boards) were not as developed at that time compared to the present day, but the internet was growing rapidly as a source of information. Second, how should the decision be made over whether a media report is positive or negative, especially as the same report could have elements of both? In earlier iterations of this indicator, the team opted for a grading system (A, B, etc.) akin to the Likert scale with one end of the scale representing "entirely positive" and the other "entirely negative", with the indicator being the median of these grades. While some guidance could no doubt be framed, there is clearly much subjectivity here as to the grading of media reports. The extremes of may be relatively straightforward but what about the grades between those extremes? But, in fairness, this is the same issue at play with the various Likert-scale type of approaches discussed in Chapter 3. Indeed, it seems reasonable to assume that the authors of the media reports may have their perceptions framed by personal experience or that of friends and colleagues as well as published values for other indicators in Table 4.2. Indeed, there could be significant influences between this perception-based indicator and the others in Table 4.2, with perception feeding upon perception. This raises the same type of concern noted in Chapter 3 about potential redundancy of indicators in any framework arising from collinearity; if the perception-based indicator in Table 4.2 can potentially be influenced by the others (via media reports), then is it necessary? Nonetheless, the group felt that while this was a potential concern and despite it being a challenging indicator to populate, it really did matter as the media had much power to influence what people thought about tourism and health; hence, its importance trumped all these methodological concerns.

To what extent does perception permeate the indicators listed in Table 4.2 beyond the assessment of media reports? Many of the indicators are not based on perception and indeed are quite standard for this theme of tourism and health. Cases of gastroenteritis can be counted and expressed on a per annum and per population basis, and similarly, there are well-established procedures for assessing the concentrations of faecal coliforms in water and pest populations via the use of traps. None of these indicators would appear to have any element of perception. Others in Table 4.2 would seem to be more subjective, even if criteria are provided as a guide for those making the assessment. There is an element of internal circularity as some indicators extend across others. Faecal coliform counts, for example, are also included in the Blue Flag system, and there is significant overlap between the Blue Flag system and the Maltese government system of grading beach quality in terms of health. The inclusion of both the Blue Flag and Maltese systems suggest the group felt it necessary to address local and international systems, and maybe this reflected the context of a period of transition for Malta as it joined the EU. There is a web of cross-influence at play even within the ten indicators selected by the group in Table 4.2. Also, it is as well to remember that the choice of the suite of indicators to capture the state of "tourism and health" in Malta, as opposed from any other suite of possible indicators, represents the view of the group as to how the theme can best be represented as a part of a much bigger picture of sustainable development of the island. The indicators are an outcome of their choices that were influenced by the discussions within the group but also by the wider contexts in play at the time. They felt it important to not only include an assessment of rodent populations but also people's encounters with them as evidenced by the number of complaints indicator. They also felt that media reporting of tourism and health, even if this was framed in part by other indicators in the list along with personal experiences and thus would itself be a perception, was important to capture as this could potentially influence people's perceptions; perceptions framing perceptions. The SSM approach to participation provided freedom for the group to think outside the box, and they took it, even if some of the indicators remained firmly within known boxes.

4.6 Challenges of participatory approaches: Capture, representation and dismissal

There is a wide spectrum of participatory approaches, each having their own advantages and disadvantages, and none of them are "ideal" for all circumstances. They differ in terms of the power vested in the facilitators to guide the process, but the outcomes still depend on choice of participants, representation and dynamics such as capture by one or a few individuals. For example, a brief comparison of focus groups and soft systems is provided as Table 4.3. Focus groups are researcher-led in almost every sense of the term, including the questions that form the basis for discussion as well as how the findings are recorded, analysed and utilised. Hence, there is potential for bias (conscious or unconscious) on the part of the researcher who can steer discussions in a certain direction, and much of potential relevance may be missed. But the advantage of being researcher-led is that the focus group addresses what the researcher needs in a relatively efficient way. Soft systems are much more participant-led, with the facilitator acting more as a guide through the steps. The result is a process that may be slower with lots of discussion, and it could potentially drift into all kinds of avenues that the researcher may not have anticipated. But in both these approaches, there remains the danger of capture by some participants while others are cajoled into staying quiet or agreeing. Even without a dominant individual it could well happen that a group just "goes with the flow", and the result may be "groupthink".

It goes without saying that with any participatory process much may depend on who is included or indeed who is excluded. The outcomes of the workshop will thus reflect the experience, expertise, motivation, worldviews and so on of those who take part. Even with the student example above, where there is relative homogeneity both within and between groups in terms of their ages, level of education and living on the same campus, there was always variation in the analysis that emerged even if similar themes (education, finance, transport, energy,

TABLE 4.3 Comparison of focus groups and soft systems approaches to participation

Focus groups	*Soft Systems Method*
Researcher-led process	Participant-led process (within the initial context set)
Participants respond to researcher questions but some freedom to explore	Participants have freedom to take the analysis in any direction they wish
Researcher does the analysis of the focus group discussions	Participants do the analysis, but researcher can compare/contrast outputs (meta-analysis)
Danger of researcher bias	Possibility of researcher bias is more limited; researcher listens
Researcher decides on direction (what happens next)	Participants decide on direction—they have control

etc.) tended to appear. It may be possible to explore differences and similarities in such group-based assessments, and Bell and Morse (2012) attempted just that for a series of SSM workshops as part of an EU-funded research project. Indeed, one intriguing facet of the SSM approach is its use of RPs at the start of the process, and this raises some potential echoes back to the use of frameworks for appraising art covered in Chapter 3. There have been efforts to apply those art appraisal frameworks to the analysis of RPs (Bell et al., 2016a, 2016b, 2019), and this remains a relatively underexplored field of study with much potential for further exploration.

One further issue with participation is linked to what happens after the process has been completed. It is all very well asking people to take part in participatory workshops and share their views, knowledge, experience and insights, but where does it all lead? Do the outputs from the participatory workshop, even when used to help plan interventions such as the SSM examples given here, bring about a positive change? Do those that took part in the process see benefits from doing so? This is potentially one of the dangers inherent in the use of all participatory approaches. It is all very well claiming that the process has helped to identify what needs to be done but what if there are no resources in place to help achieve any of this? Identifying interventions that the community can take themselves without any outside support may have positive impact and that should not be undervalued, but what if the community identifies a need for support from others such as government who then ignore them? Participation may result in all kinds of expensive "wishes", but there are no guarantees that those with the power to do something will do so, even if they sponsored the participation in the first place. Understandably, the result can be frustration and a "participation fatigue" may set in where communities feel that undergoing a participatory exercise achieves nothing of any benefit to them (Morse, 2008). Even worse, we may see the rise of a kind of "methodological tyranny" where participatory approaches, even those using approaches that ostensibly sit towards the top end of Arnstein's Ladder, are undertaken almost for the sake of it, and in reality, the approach is really one, at best, of consultation (Morse, 2008; Morse and Bell, 2011).

As shown with the examples above, participatory approaches can be used to identify and even develop indicators to help with attainment of sustainability. Once a set of interventions have been identified, along with answers to the BITOAC checklist, it is possible if not desirable to set out indicators to help capture the sought transformation and its effectiveness. Hence, indicators can be an important part of the monitoring and evaluation of interventions. But there is more to indicators than just an instrumental use linked to interventions. Sustainability is complex, and participants in a participatory process may find it challenging to define what needs to be done, by whom and how. Expressing all of this in terms of indicators, such as those for "tourism and health" in Table 4.2, can help participants focus on what they see as their priorities and allow them to develop and indeed frame their own insights (Morse, 2008), such as the need to assess media reporting. Hence, indicators may help facilitate debate and raise awareness and thus dovetail neatly with participatory processes which can also help facilitate learning between participants (Bell and Morse, 2003; Morse, 2008).

4.7 Summary

The generation of new knowledge via research is typically an extractive process—the researcher surveys or interviews people to glean answers to questions that the researcher has set, and it is the researcher who controls the analysis and identification and expression of any recommendations. People may well be a part of this process, but their role is the provision on information required by the researcher; they have little if any control over the process. Participatory methods are designed to involve people far more in the decision-making process—in effect to hand more power to those who are engaged in the process, so they can help set questions, provide answers and have an input into what should happen next. But there are many ways of engaging people, and Arnstein's Ladder sets out a spectrum of approaches from manipulation at the bottom of the ladder to citizen control at the top. There exists a plethora of different ways of "doing" participation, all of which can be placed somewhere along the Arnstein framework as shown by Geekiyanage et al. (2021), and each of them has its own set of advantages and disadvantages. It is important to recognise that the Arnstein ladder was never meant to be a scale representing an intrinsic "value" of the approaches, with a possible exception of manipulation which does have a somewhat nefarious feel. They all have their own use depending upon context.

The chapter sets out two approaches to participation, focus groups and SSM, with an especial emphasis on the latter, and how it has been applied in practice to help developed SIs. All participatory methods can make use of indicators to help explain and articulate issues, but they can also be used to identify and develop appropriate indicators. Indeed, it can perhaps be hypothesised that participatory methods can naturally lead to the development of indicators based on perception. When people are handed control over the process of indicator development, it seems probably that they may take a broader perspective and think outside the "box". The "tourism and health" set of indicators developed using SSM had a mix of what can be regarded as well-established indicators along with some indicators that are more subjective but also regarded as important given that they can influence the perceptions of others. The inclusion of an indicator intended to assess media reporting of tourism and health is an intriguing example given that the media may not only influence the perceptions of readers but may in turn be influenced by indicators. But, in fairness, this more "perception friendly" hypothesis for developing indicators via participatory approaches, at least when compared to more "top down" approaches such as relying on teams of experts, remains just a hypothesis.

Participatory approaches are not a panacea and should not be considered as necessarily the only or even necessarily the best way to involve people in the generation of insights and ideas for problem solving. There is no intention in this chapter to reify participation as the best way to achieve goals; that would be a mistake. Approaches founded on questionnaire-based surveys and semi-structured interviews also have their place.

4.8 Food for thought

1. Please derive other ways of categorising participatory methods that build on the Arnstein Ladder.
2. What are the challenges involved in the use of participation compared to other approaches to research such as questionnaire-based surveys?
3. Is it necessary to consider group dynamics when it comes to participation and how can they be taken into account?
4. How can analytical frameworks designed to guide the appreciation of art be applied to RPs?
5. How would you test the hypothesis that participatory approaches lead to more perception-based indicators compared to more expert-driven approaches?

References

Adelman C (1993) Kurt Lewin and the Origins of Action Research. *Educational Action Research* 1(1), 7–24

Arnstein SP (1969). A Ladder of Citizen Participation. *Journal of the American Institute of Planners* 35(4), 216–224

Bell S and Morse S (2003). *Measuring Sustainability. Learning by Doing*. Earthscan, London

Bell S and Morse S (2008). *Sustainability Indicators. Measuring the Immeasurable*. 2nd edition. Earthscan, London

Bell S and Morse S (2012). *Resilient Participation: Saving the Human Project?* Routledge, Abingdon

Bell S and Morse S (2014). Groups and Indicators in Post-Industrial Society. *Sustainable Development* 22(3), 145–157

Bell S, Berg T and Morse S (2016a). *Rich Pictures: Encouraging Resilient Communities (Earthscan Tools for Community Planning)*. Routledge, London

Bell S, Berg T and Morse S (2016b). Rich Pictures: Sustainable Development and Stakeholders—The Benefits of Content Analysis. *Sustainable Development* 24(2), 136–148

Bell S, Berg T and Morse S (2019). Towards an Understanding of Rich Picture Interpretation. *Systemic Practice and Action Research* 32(6), 601–614

Benor D and Harrison JQ (1978). *Agricultural Extension: The Training and Visit System*. World Bank, Washington DC

Bolton K and Brace I (2022). *Questionnaire Design: How to Plan, Structure and Write Survey Material for Effective Market Research*. Kogan Page, London, New York and New Delhi

Chambers R (1983). *Rural Development. Putting the Last First*. Prentice Hall, Harlow

Chambers R (1994). The Origins and Practice of Participatory Rural Appraisal. *World Development* 22(7), 953–969

Chambers R (1997). *Whose Reality Counts? Putting the First Last*. Intermediate Technology, London

Chambers R (2017). *Can We Know Better? Reflections for Development*. Practical Action Publishing, Rugby, UK

Checkland PB (1998). *Systems Thinking. Systems Practice*. John Wiley & Sons Ltd

Checkland PB (2000). Soft Systems Methodology: A Thirty-Year Retrospective. *Systems Research and Behavioral Science* 17, 11–58

Checkland PB and Poulter J (2006) *Learning for Action: A Short Definitive Account of Soft Systems Methodology and Its Use for Practitioners, Teachers and Students*. Wiley, Chichester

Checkland PB and Scholes J (1999). *Soft Systems Methodology in Action*. Wiley, Chichester

Freire P (1976). *Education, the Practice of Freedom*. Writers and Readers Publishing Cooperative

Freire P (2000). *Pedagogy of the Oppressed* (30th anniversary edition). Bloomsbury, New York

Geekiyanage D, Fernando T and Keraminiyage K (2021). Mapping Participatory Methods in the Urban Development Process: A Systematic Review and Case-Based Evidence Analysis. *Sustainability* 13(16), 8992. https://doi.org/10.3390/su13168992

Goldhill S (2011). 'In Our Time'. *BBC Radio 4*, May 26, 2011. Available at: www.bbc.co.uk/programmes/b011cffd

Hesser L (2006). *The Man Who Fed the World: Nobel Peace Prize Laureate Norman Borlaug and His Battle to End World Hunger*. Durban House Publishing Company, Dallas, USA

Jain HK (2018). *The Green Revolution: History, Impact and Future*. SP LLC, USA

Lewin K (1946) Action Research and Minority Problems. *Journal of Social Issues* 2(4), 34–46

Lewin K (1948). *Resolving Social Conflicts, Selected Papers on Group Dynamics (1935–1946)*. Harper, New York

Lewin K (1958). *Group Decision and Social Change*. Holt, Rinehart and Winston, New York

Morse S (2008). Post-Sustainable Development. *Sustainable Development* 16(5), 341–352

Morse S (2015). Developing Sustainability Indicators and Indices. *Sustainable Development* 23(2), 84–95

Morse S and Bell S (2011). Sustainable Development Indicators: The Tyranny of Methodology Revisited. *Consilience: The Journal of Sustainable Development* 6(1), 222–239

Schugurensky D (1998). The Legacy of Paulo Freire: A Critical Review of His Contributions. *Convergence* 31(1/2), 17–29

Shiva V (1991). *The Violence of the Green Revolution: Third World Agriculture, Ecology and Politics*. Zed Books, London

Talubo JP, Malena RA, Morse S and Saroj D (2022). Practitioners' Participatory Development of Indicators for Island Community Resilience to Disasters. *Sustainability* 14(7), 4102

Truman HS (1949). Inaugural Address. January 20, 1949. Available at: https://www.trumanlibrary.gov/library/public-papers/19/inaugural-address

Xenophon (2009). *Anabasis: The Persian Expedition*. CreateSpace Independent Publishing Platform

Yurick S (1965). *The Warriors*. Holt, Rhinehart & Winston, New York

5

COMMUNICATING AND USING PERCEPTION-BASED INDICATORS OF SUSTAINABILITY

5.1 Introduction

It was suggested in an earlier chapter that much of the emphasis within indicator research has focussed on the technical development of indicators, which typically spans making a case as to why a new or a modification of an existing indicator is required, and the methodology behind it. In addition to this, there is an abundance of research that uses these indicators to make comparisons between countries, maybe also linked to trends and patterns over time. Arguably, there has been far less attention on how indicators are used, by whom and for what purpose. There is often a claim that indicators will improve matters as "policy makers", and other stakeholders can "use" them to help guide decisions, but there has been surprisingly little attention given to whether such target groups do indeed "use" the information let alone, how they do it and what changes are brought about as a result. Given that indicators created to help bring about a desired change, this may seem like an odd gap, but there are reasons why such an exploration is fraught with challenges, and these will be outlined later in the chapter.

Nonetheless, a key stage in the use of any indicator is its communication, and one of the key groups typically selected as an aid in this task is the media. The logic here is clear and compelling; the traditional media (newspapers, TV, radio, etc.) can provide a vehicle for communicating the indicator and associated insights to a wider audience that spans the public, business and government amongst many others (Morse, 2023a, 2023b). Indeed, social media platforms provide an additional vehicle for communication, and for some demographics, it is the main vehicle for consuming news (Morse, 2023a). There are, of course, numerous bi-directional interactions in all of this, and social media platforms may well pick up on stories in the traditional media and vice versa. Also, an expectation here could be that

DOI: 10.4324/9781003456605-5

the public, when faced with the insights provided by the indicators, can put pressure on others (government, business, etc.) to act, and the same may be true of all stakeholders in this web of communication. The interplays of communication and influence are potentially numerous and complex, but communication, especially to the media, does seem to be an important first step which underpins use, and that is why, it is the primary focus in this chapter.

This chapter will focus on the reporting of indicators via the traditional and social media as a form of communication but also of use of the indicators by the media. This is, admittedly, a significant limitation, but the logic behind it is that any indicator must first be communicated to those who are most likely to use it to help make decisions. In many cases, of course, the indicator is communicated via a set of documents typically comprising a main report with all the detail, sometimes including technical information on the methodology used for the indicator, allied with a summary of the main conclusions. The summary may be framed as a "press release" or perhaps as an executive summary, or maybe both. Before the advent of the internet, the material was circulated in "hard" form to the media, government officials and other "stakeholders" who may be interested, but these days, the communication is much more via websites and associated electronic mailing list and social media posts. Nonetheless, while the mechanics of the communication has changed significantly over the years, the assumptions and motivation behind indicators remain the same; there is little point in producing an indicator if no one has access to it and act upon the message(s) it conveys. But the media do not solely act as a vehicle for communication; they are also free to use the indicators in their reporting in any way that they wish. Hence, when exploring the reporting of indicators by the media, there are these two facets to consider—one is communication (and hence raising awareness of the indicator and the issues it is linked to) and the use of the indicator by the media to tell the stories it wishes to tell. These two facets are clearly related, and indeed, it is not always possible to see a dividing line between communication and use.

The chapter begins with an outline as to what the academic literature tells us about the main types of use that are often ascribed to indicators. This short review is followed by sections, which explore the extent of reporting of indicators by newspapers and social media. Questions covered here include how indicators are used by reporters and what they use them for? Of especial interest is whether there are any differences in the extent of reporting of perception-based and non-perception-based indicators within the newspaper media. The latter part of the chapter moves on to explore newspaper reporting of the CPI as it relates to two countries—Nigeria and Australia—and differences between the reporting.

5.2 Who uses perception-based SIs and why?

We may wish, indeed like, to think that policy development and indeed the design and implementation of interventions are all solidly evidence-based and that the interventions are designed based on knowledge of what best delivers the intended outcome. In other words, those in charge have evidence that they can draw upon, and indeed do draw upon, to make desired change happen as intended (Neylan, 2008; Witting, 2017). The term "evidence-based" provides a sense of objectivity; a picture that evidence has been derived objectively and has been applied to bring about a change where we will have confidence in what happens. It provides a level of comfort and reassurance that we will get what we want to get rather than no change at all or unintended consequences that may make matters worse. But, of course, for such a neat cause–effect to be the outcome much depends on the availability of evidence and an ability by those charged with making these decisions to find it and absorb it in such a way that helps them come to their decisions (Huston, 2008; Hallsworth et al., 2011; Cairney and Oliver, 2017; Witting, 2017). Indeed, plenty of evidence could well be available, but it may point us in quite different, perhaps contradictory, directions. An example is provided in Chapter 3 with the various indicators used for assessing crime. One indicator based on crimes reported to police may suggest a decline in the crime rate while another indicator based on people's experiences and perhaps perceptions of crime may point in the opposite direction, which suggests that matters are becoming worse. Which of the stories painted by the two indicators, or indeed many other sources of evidence that may be available, should we believe? This is a common occurrence, and decision-making can typically be subject to what is called "bounded rationality" where, for example, in the case of policy development:

> In part, bounded rationality relates to the fact that policymakers do not have the ability to gather and consider all evidence relevant to policy problems. Instead, they employ two shortcuts—"rational", pursuing clear goals and prioritising certain sources of information, and "irrational", drawing on emotions, gut feelings, beliefs and habits to make decisions quickly.
>
> *(Cairney and Oliver, 2017; page 3)*

Hence, "bounded" is used to note that there are bounds or limits to "rationality" (i.e., decisions based on evidence). Just as important, maybe more important in some cases are the irrational (i.e., not based on evidence) factors. The latter, of course, can also have significant political over- and undertones as, after all, civil servants and managers may have to work within a broad context provided by politicians. Evidence may point in one direction, or perhaps no clear direction at all, but political dogma or perhaps the feedback that politicians are receiving from their constituents may trump this evidence and nudge them to take a different set of decisions. Welcome to the messy world of decision-making, where decisions can be subject to a host of influences including pressure from certain groups and organisations in society that have their own interests at heart. Indicators, of all types, are often part and parcel of these debates and influences, and groups may quote and wield their own choices of "facts" backed up by indicators to support their case and discredit alternative visions created by their opponents. There are no guarantees that the "best" indicators are the ones that win the day or that everyone will agree on what the "best" indicators are (Turnhout et al., 2007). Indicators, by their very nature and intention, can end up being a part of these messy debates and can generate more heat than light.

While the literature on the use of indicators is not as extensive as I would like, the studies that do exist suggest they have various uses. Indeed, typologies have been around since the early 2000s that set out the various types of use of indicators, especially within the policy domain by local and national government departments and agencies (Gudmundsson, 2003; Eckerberg and Mineur. 2003; Hezri, 2004, 2005; Hezri and Hasan, 2004; Hezri and Dovers, 2006; Ramos and Caeiro, 2010; Lyytimäki et al., 2013; Rinne et al., 2013). These frameworks vary in terms of their focus and complexity, but they do have points in common. For example, the Hezri (2004) typology has two main types of indicator "usage", summarised as follows with some indicative first-person statements provided as illustration:

- Instrumental use: Here, it is assumed that there is a direct link between the values of indicators/indices and decision outcomes; "if the indicator has the value of xxx then we should do yyyyy"
- Conceptual use: In this case, the indicators/indices may act as catalysts to sensitise/change people's understanding of a problem or situation; "Economic productivity is a complex issue for me to grasp but explaining it as economic output (GDP) generated by each person working helps me to see what you are getting at"

Both these are, in a sense, "constructive" uses of indicators in that they are being used to help inform actions and to spur learning and understanding. The instrumental and conceptual uses are perhaps those most closely aligned with those typically envisaged by the creators of indicators. Indeed, "instrumental" is perhaps the use that most people have in mind for indicators in the sense that they are populated on a regular basis with good quality data and help to inform decisions. "Conceptual" use of indicators is different to "instrumental" and indeed has many echoes back to Chapter 4 on participatory methods of indicator development. Here, the indicators act as a means for framing what can be a complex issue (Rydin, 2007); put another way, they can help with learning and understanding complexity. Vague terms such as a more "sustainable society" may sound good and desirable but what is meant by "more sustainable"? For a community, it may boil down to aspects such as "more employment" and "less crime", amongst many others of course, and indicators can play a role in helping to define these in ways that are more tangible and thus meaningful. Hence, indicators can be used to help "steer" or "nudge" a decision-making process into a particular direction, although this change can take time (Morse, 2023a). The idea of "nudging" change in behaviour has been a popular one, especially amongst governments

(Thaler and Sunstein, 2008, 2021), but the potential for indicators to help with this remains relatively underexplored (Morse, 2023a). Enhanced awareness may, in turn, feed through at some time into decision-making but perhaps in ways that are incremental (Hezri and Hasan, 2004) and hence challenging to identify and pin down in any clear sense of "cause-effect". Indeed, the web of influence can be far more extensive and subtle, and the indicators can help raise awareness of issues amongst a wide range of actors all of whom can in turn bring influence to bare on those who make decisions. Dissecting such a complex and subtle web of influence of indicators over time may present quite a challenge as even the actors themselves may not necessarily be able to identify or appreciate the influence let alone judge its importance vis-à-vis the many other influences at play that have helped shape their world view.

At one level, it seems reasonable to assume that perception-based indicators are no different from other indicators in this regard, and both instrumental and conceptual types of use will apply. As discussed in Chapter 3, indicators that are based on perception often exist because it can be challenging to measure something directly and with confidence. Corruption is a good example. The CPI exists in the more indirect form that it does because it can be difficult to measure corruption in ways that are more direct. But even with this limitation, the CPI can still be employed in an instrumental sense and provide a warning that something needs to be done, even if it is not always clear what the "something" should be. Similarly, the CPI allows for cross-country comparisons as well as comparison for the same country over time, and these can aid with illustrating the scale of the problem and perhaps what causes it; an analysis that has resonance with the examples highlighted in Chapter 3 for assessing what influences happiness. The CPI also helps make the point that even the least corrupt countries in the world, those at the top of the CPI rankings, are still perceived to have some corruption; no country is perfect.

5.3 Reporting perception-based indictors by the media

Indicators are often reported in the mainstream and social media, and this is not surprising given that the media, in all its forms, is often a major target for those who have created the indicators in the first place. Looking at it from the other direction, journalists may perceive those providing indicators as being "authoritative" sources of information. For example, Lowrey and Hou (2021; page 37) have noted how:

> [J]ournalists derive legitimacy and authority from the political-economic institutions that produce most of the data sets. When this dependence is strong, it does little to encourage journalists' critical thinking. Also, quantitative accounts and data categories, and the scientific practices used to create them, have an aura of authoritative objectivity, and this appeals to beleaguered journalists and news outlets, beset with charges of bias and challenges to their authority.

Similarly, politicians, businesses and others may seek to influence the messages being conveyed by the media (Barabas and Jerit, 2009), and that can include how indicators are being interpreted as well as seeking to influence the choice of indicators by the media. But how does media coverage of perception-based indicators differ from indicators that are not perception-based? Figure 5.1 shows the number of newspaper articles published across the globe (in many languages) that mention some examples of perception-based indicators at least once. In some cases, the articles mention the same indicator a couple of times but that would still be counted as one "use". The searches were undertaken using the Nexis database of newspaper articles, and the search was undertaken across all countries but limited to newspapers rather than other types of publication. Details of the Nexis database and how searches are undertaken are provided in Morse (2023a). Each of the graphs from (a) to (g) present the trend for the indicator from 1990 to 2023.

For all the graphs in Figure 5.1, there is a discernible (albeit more apparent for some indicators than others) increase in the number of published newspaper articles from 1990 to 2023. Part of this trend is no doubt due to an increasing awareness of the existence of the indicators amongst journalists, perhaps caused by a ratcheting up of efforts by those promoting them and the rise of the internet and electronic communication tools such as email and social media. Both the internet and email have a long history that can be traced back at least to the 1960s, but their use started to become more widespread from the 1990s onwards. Hence, the increase in the number of newspaper stories from 1990 onwards may reflect in part an increase in effort and the effectiveness of that effort at promotion. But it can also be theorised that some of the increase in article count is due to an increase in perceived usefulness of indicators for making or supporting a point. Hence, and as we will see later in the chapter, in an article on the prevalence of corruption in a country or group of countries, the journalist may use the CPI to illustrate prevalence and/or trends. The article itself may not be focussed on the CPI, but the indicator is used to help make a point and provides the kind of "authoritative" support noted above by Lowrey and Hou (2021). Nonetheless, while the trends over time seen in Figure 5.1 may have various causes, what is especially interesting is the marked difference in article counts between the indicators. It seems the big "winner" in terms of media popularity in Figure 5.1 is the CPI. This is perhaps not surprising, given widespread interest in corruption, with its length of "service", the CPI has had a long life (since 1995), and the significant effort placed into its dissemination by TI. But I suspect that of these explanations, it is the nature of the issue being covered, corruption and the sensitivities (including political) that surround that issue which explain much of the popularity of the CPI within this reporting. The other three perception-based indices that appear to have some degree of resonance in newspapers are the "Happiness Index", Net Promoter Score and Democracy Score. The "Happiness Index" is an interesting one and has been covered in depth in Chapter 3, and its popularity amongst journalists is perhaps not all that surprising; after all, we are all interested in happiness! Of the other indicators in Figure 5.1 (Economic Confidence Index,

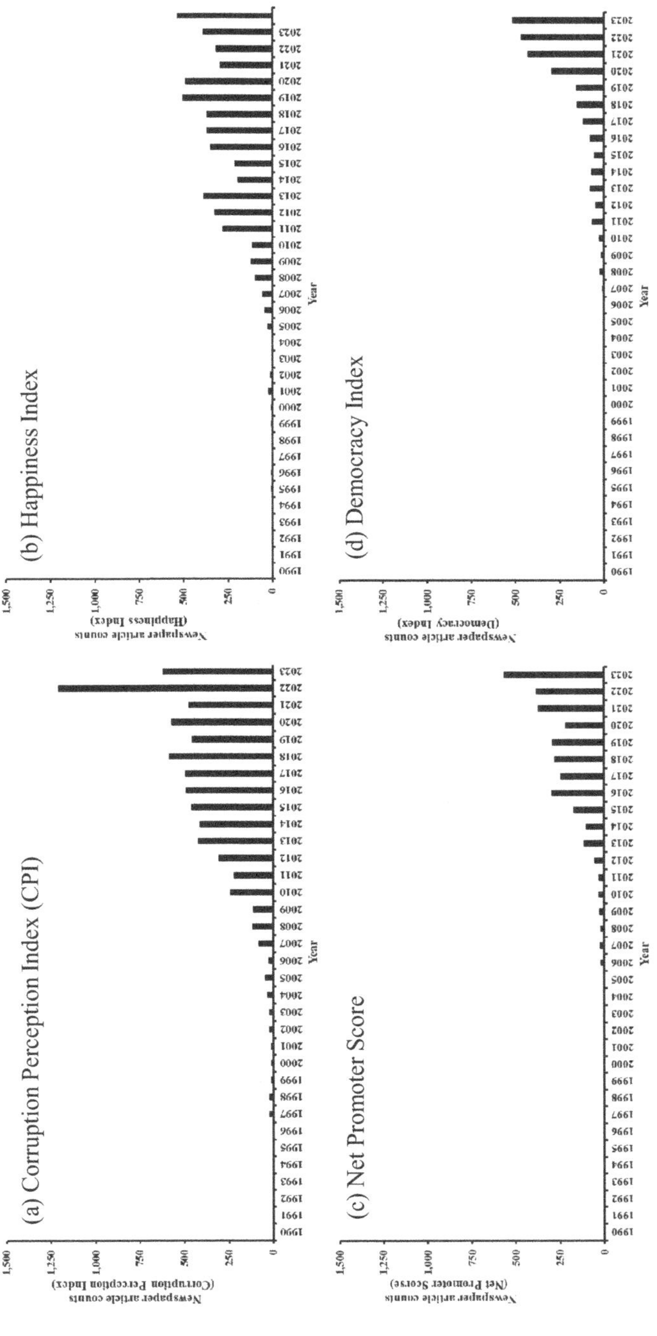

FIGURE 5.1 Newspaper reporting of some perception-based indicators between 1990 and 2023

The Figures shown in the graphs are the number of articles published each year that mention the indicator at least once.

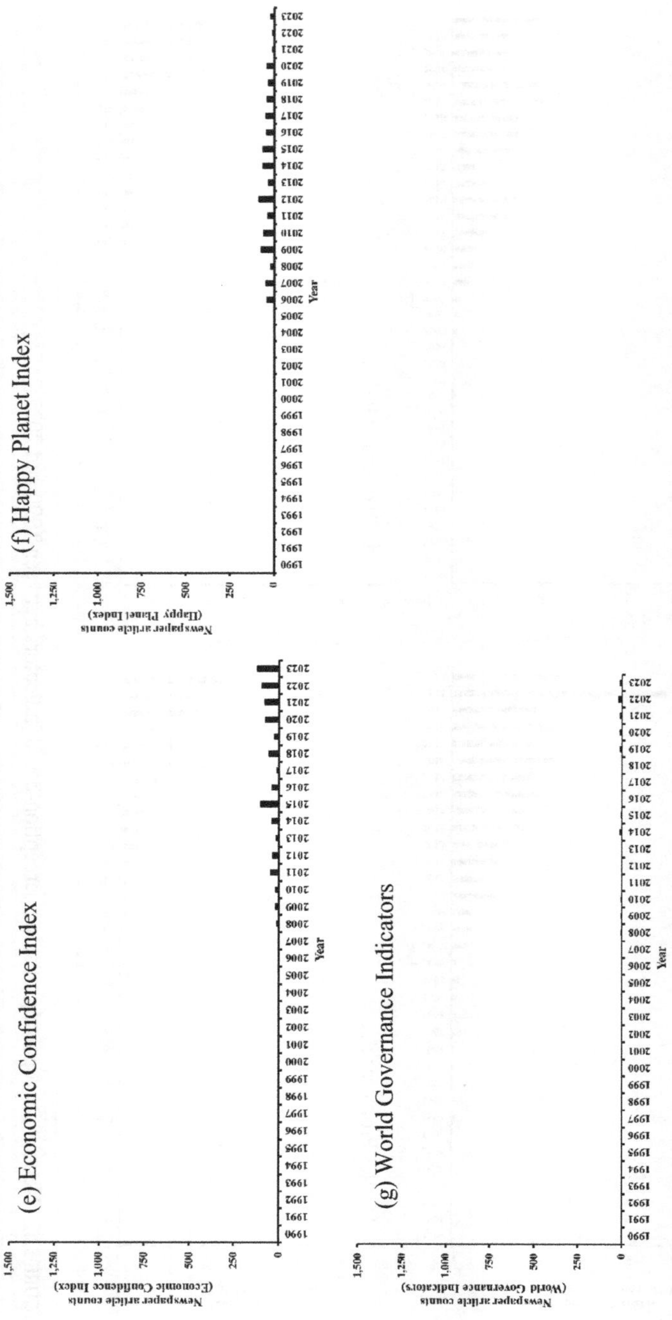

FIGURE 5.1 (Continued)

Happy Planet Index, WGI), the number of newspaper articles that mention them is very low; indeed, in the case of the WGI produced by the World Bank, the numbers hardly register on the graph; you may need a magnifying glass to see the bars!

5.4 Reporting non-perception-based indictors by the media

How do the counts of newspaper articles for perception-based indicators compare with those that are not perception-based? There are many indicators that are populated based on data collected via surveys and other sources which are not based on eliciting perceptions, and a few of them are shown in Figure 5.2. The first graph (Figure 5.2a) is newspaper article counts for the HDI; a widely reported indicator created by the United Nations Development Programme (UNDP) and published (almost) annually since 1990. The HDI is often used as a proxy measure for "quality of life" and "wellbeing" by the media although neither of these was intended by its creators (Morse, 2023b). The HDI is an amalgam of three indicators covering education (enrolment rates in schools, etc.), health (proxied by average life expectancy in years) and income (GDP/capita; with GDP adjusted for purchasing power). The HDI does not include any dimension for crime, happiness or indeed the environment, and given that GDP (later the GNI) is used as the basis for measuring income, then all the limitations of that indicator discussed in an earlier chapter apply here. But the HDI does have the important advantage of being promoted by the United Nations, an extensive international organisation having many offices worldwide. Whenever the latest version of the HDI is released, it is accompanied by much fanfare and publicity, with much of the latter targeted at the media. Hence, it is perhaps of no great surprise that while the pattern in newspaper reporting of the HDI is broadly in line with that for some of the perception-based indicators in Figure 5.1, with a maximum around 1500 articles/year, the number of articles that mention the HDI is significantly larger.

The second non-perception-based indicator (Figure 5.2b) is the EF, which is arguably one of the more widely reported indicators in the media having a specific focus on the environment. The EF is a measure ("footprint") of how much impact our consumption of "stuff" has on the planet; the higher the EF the greater the size of the footprint (i.e., impact). The notion of representing our impacts on the environment via a "footprint" is a powerful one, and while the newspaper reporting of the EF is well below that for the HDI, it is still higher than some of the perception-based indicators in Figure 5.1. Incidentally, the notion of a "carbon footprint" (i.e., the global warming potential of GHGs) sounds similar but is in fact quite different to the EF in terms of its conceptualisation and calculation although the EF does include a global warming component as part of its impact (Morse, 2019).

The most widely reported indicator by orders of magnitude in Figure 5.2 is the GDP (see Figure 5.2c). The reporting of this economic indicator dwarfs all the others in Figures 5.1 and 5.2. Indeed, the number of articles that report the GDP is far

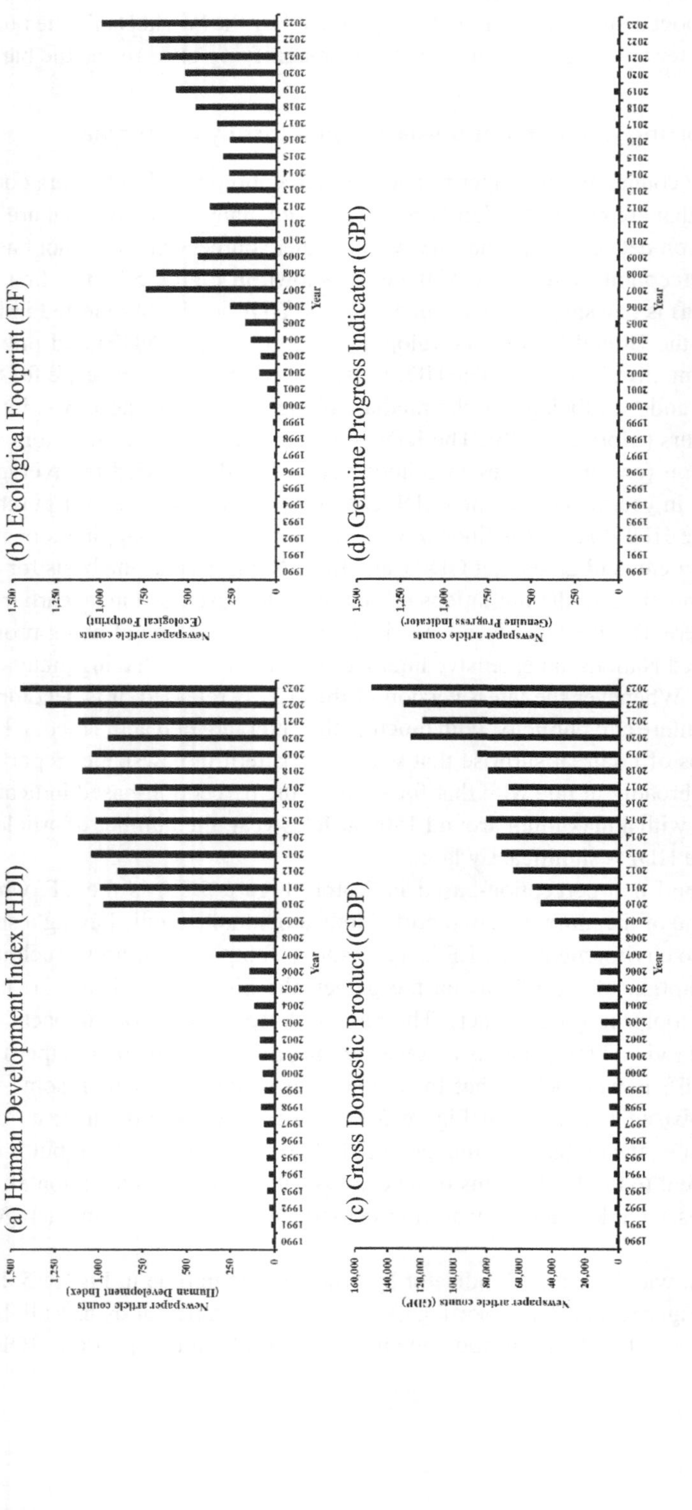

FIGURE 5.2 Newspaper reporting of some indicators that are not based on perception between 1990 and 2023

The Figures shown in the graphs are the number of articles published each year that mention the indicator at least once.

greater than the sum of all the other indicators, perception and non-perception-based, as shown in Figures 5.1 and 5.2. The popularity of the GDP in the media is readily understood given the emphasis given to economic performance by politicians, businesses and indeed governments. Indeed, if anything, the dominance of such economic indicators may perhaps be under-represented here, as there are "variants" on the GDP theme (e.g., GNI, GNP), which were not included in the search. There have been attempts to re-formulate the GDP to include expenditures on activities deemed to be good for society such as volunteering and caring which do not involve payment (i.e., flow of money). Indeed, it is possible to boost GDP by degrading the environment and having very high levels of crime (hence, the need to spend money on crime prevention), but these are not "desirable" attributes in society. The GPI is one such attempt to balance the GDP by including desirables and removing undesirables. The immediate precursor of the GPI was the ISEW developed in the 1980s with the GPI emerging some years later in the mid-1990s. The GPI has thus been around for the same length of time as the CPI and is only 5 years or so younger than the first publication of the HDI, but the GPI has not received anything like the same attention in newspapers as these indicators (Figure 5.2d). Looks like you will need that magnifying glass again to see the bars!

5.5 Multiple pictures of nuance: Reporting the CPI

As an illustration of the different uses of one of the perception-based indicators in the media, notably the CPI, and indeed to add more nuance to the simple counts of newspaper articles in Figure 5.1a, I have opted to focus on newspaper reporting that makes use of this indicator for two countries that are very much at opposite ends of the rankings—Australia and Nigeria. The aim here is to go into more depth and provide illustrations of the ways in which the CPI is used within the newspaper stories: What are the contexts for its use? What kind of associations or trends is it used to illustrate? What points are being made and to what purpose?

Australia sits towards the top end of the CPI country rankings (i.e., the end of the ranking that equates to being less corrupt), and since 1995, the country has, on average, ranked in the top ten of least corrupt countries of those included in the CPI reports published by TI. Countries that have ranked number one (the least corrupt of all those included) has fluctuated between New Zealand, Denmark, Finland and Iceland, but Australia has consistently been up there amongst those ranked as least corrupt. The other country selected here for attention is Nigeria, which has tended to sit at the other end of the rankings than Australia. Nigeria has often been ranked as the most corrupt country in the CPI "league table", and indeed, the country sat at the foot of the CPI rankings in 1996, 1997 and 2000. Other countries that have occupied the unenviable position of being "most corrupt" in the CPI rankings include Cameroon, Bangladesh, Haiti and Somalia. Australia and Nigeria are not especially connected in terms of trade. The bulk of imports (99%) Australia has from Nigeria are petroleum products, and in the other direction, the main export from Australia to Nigeria are delivery trucks and cars, and the trade volumes are not especially high (Observatory of Economic Complexity, 2025). Hence, the choice of these two countries for a more in-depth analysis of newspaper reporting of the CPI is not solely based on their opposite rankings but also because they are not connected significantly via trade, and thus, the reporting is unlikely to have much, if any, overlap. There is an argument that can be made, of course, to do the same kind of analysis for countries that sit at the top and bottom of the rankings that are much more intertwined in terms of trade, perhaps because they could bring out different perspectives on how corruption affects that relationship.

The counts of articles published each year for Nigeria and Australia that mention the CPI at least once are shown in Figure 5.3. As mentioned above, few of these stories mention both Nigeria and Australia so the overlap between these counts is minimal. The numbers in Figure 5.3 tell their own story as they are considerably

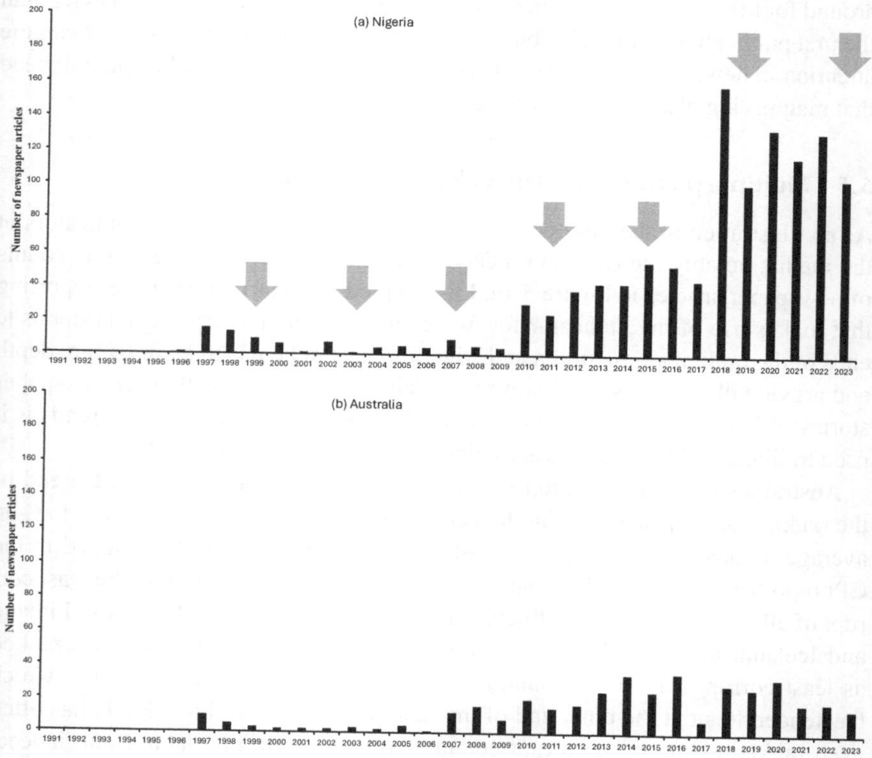

FIGURE 5.3 Number of newspaper articles published each year that mention the CPI at least once

Data for two countries are provided here (a) Nigeria and (b) Australia. For Nigeria, the arrows mark the years when presidential elections were held.

higher for Nigeria than Australia. This may not be all that surprising as there is an old adage in media circles that bad news tends to travel far and fast, and indeed, various academic studies have confirmed that the media can often have what is called a "negatively" and "bad news" bias (Haskins and Miller, 1984; Soroka, 2006; Garz, 2014; Kätsyri et al., 2016; Kollmeyer, 2004). The relatively high level of corruption in Nigeria is certainly "bad" news for the country and its people, and it is not hard to see why it would attract much attention from the media both within Nigeria and outside. But there may be more to it than a desire on the part of journalists to cover this topic for the public good; there can also be a desire to sell stories. Aday (2010; page 146) has noted that "*one reason for the persistence of a negativity bias across news genres can be traced to the fact that conflict is what Herbert Gans (1980) found to be a fundamental news norm, something that journalists look for in defining a story as newsworthy. "Good news" is generally seen as less interesting to viewers than "bad news", especially on television*". In essence, sensationalism, magnifying negative and positive aspects of a story, can simply sell more copies of newspapers (McCluskey et al., 2016). It may seem unfair, but the high level of corruption in Nigeria is more "newsworthy" for want of a better term than the good news story of less corruption in Australia.

In Figure 5.3, we can see that the distribution of counts for both countries is not evenly spread, and there are some years when the number of stories seems to surge. Given that the number of articles is notably higher for Nigeria than Australia, the changes from year-to-year are more pronounced, and one possibility is that the media interest in the CPI may be linked to election cycles. While corruption in Nigeria is a continuous theme in newspaper reporting, it is after all commonplace in the country, and there are many victims at all levels in society, and it is reasonable to assume that it could become an especially relevant issue during times of electioneering. Politicians often make promises to reduce or even eliminate corruption, and the track records of incumbents are often held up to scrutiny. In Figure 5.3a, the arrows represent years of presidential elections, and the years prior to the election often show a heightened level of reporting of the CPI by newspapers.

While the numbers in Figure 5.3 tell us something about the trends in newspaper reporting that makes use of the CPI, it is also of interest to provide some context. I cannot present every one of the newspaper stories behind the counts in Figure 5.3, but I can at least provide the reader with a taste of that reporting. Hence, I have included excerpts (quotations) from some of the stories with my own highlights (bold and underline) of key points. I have also grouped them into themes although I am in no way trying to suggest that this is based on an exhaustive content analysis of the material. My aim here is simply to provide some richer context to the simple counts of newspaper articles and provide a sense as to how the CPI is being used. Some of the quotes are quite lengthy in order to preserve context as much as possible, and all emphases (bold text and underlined) have been added by me. I have included some clarifications using square brackets.

The first set of quotes below have been taken from articles published in Nigerian newspapers between 2018 and 2023; years that have the highest counts in Figure 5.3a. A presidential election was held in February 2019 (won by Muhammadu Buhari with approximately 56% of the votes cast) and another in February 2023 (won by Bola Tinubu with approximately 38% of the votes cast), and in both cases, there were primary elections within the political parties the year before. The election results were contested by the losing candidate, as they often are in Nigeria, but eventually they were upheld by the courts. The causes, extent and control of corruption often emerge as a significant issue in the election campaigns.

Theme: Extent of corruption and failure of governance

> Quote 1: Corruption is getting worse in Nigeria, according to the latest corruption perception index (CPI) released by Transparency International (TI) on Wednesday. While the country scored 27/100 and was ranked 136th in 2016, the latest CPI scores Nigeria 28/100 but with a rank of No. 148 out of 180 countries surveyed—a significant 12 places below where it was the previous year. *This will come as a blow to the President Muhammadu Buhari administration who came into office on the strength of his anti-corruption credential.*

Title: Corruption Getting Worse in Nigeria—TI
Source: Vanguard (Lagos), 21 February 2018
Author: Anonymous

> Quote 2: Many of those who voted for Buhari in 2015 genuinely believed he was going to put an end to corruption or at least tackle it. According to Transparency International's latest Corruption Perception Index, Nigeria ranks 146 out of 180 in its Index, slipping 10 places from 136th place when Buhari was first inaugurated in 2015. Thus, *under Buhari, corruption has not reduced but runs rampant, expanding in size and intensity.*

Title: Buhari's Anti-Corruption War and the Economy
Source: This Day (Lagos), 29 November 2020
Author: Nosa James-Igbinadolor

> Quote 3: On TI's annual Corruption Perception Index, which ranked Nigeria 148 out of 180 countries assessed in 2017, showing deterioration in perception of corruption in public administration in Nigeria compared to 2016, Governor Fayose said: "Transparency International has placed a *question mark on the president's claim to integrity.* Fact is that President Buhari is presiding over the *worse form of corruption* in the history of Nigeria and the good thing is that despite their propaganda, they have *not been able hide the rot in their government* from the eagle eyes of international organisations like the Transparency International".

Title: Yobe Schoolgirls' Abduction Exposes Govt's Lies On Boko Haram—Fayose
Source: This Day (Lagos), 23 February 2018
Author: Victor Ogunje

> Quote 4: Atiku took a swipe at Buhari's avowed war against corruption, and noted that the country has recorded its worst in Corruption Perception Index (CPI) in recent years. He said, "In 2018, Transparency International announced that Nigeria made her worst ever retrogression in the Corruption Perception Index moving 12 steps backwards from 136 under the Peoples Democratic Party to 144 under Buhari. *Nigerians are asking if this administration is planning 'next level' corruption for them?*

Title: Atiku Mocks Buhari's "Next Level" Campaign
Source: Daily Independent (Nigeria), 19 November 2018
Author: Samuel Ogidan

> Quote 5: Former President Goodluck Jonathan as formally responded to the infamous claim that he tended to downplay raging corruption in his administration by boldly claiming that *'stealing is not corruption'*. But in dispelling the claim, Jonathan claimed that corruption was more prevalent under the President Muhammadu Buhari administration than under him and *dismissed the current anti-corruption fight as mere media show*. The former president boasted: "It is important to note that despite the many sensational stories, dramatic arrests, seizures and accusations, many of them false, since I left office, the fact remains that Nigeria has not made any improvement on TI Corruption Perception Index since 2014. In fact, the 2017 CPI released in 2018 by TI places Nigeria as number 148, a retrogression in which the nation went 12 place backward. In other words, Nigeria is more corrupt in 2017 than it was when I handed over to Buhari administration in 2015. Some people may be misled with smoke and mirrors but the *TI Corruption Perception Index relies on unsentimental facts and figures*", he submitted in this book, which seeks to set the records of his administration straight and debunk false narratives by commentators against his tenure.

Title: Nigeria More Corrupt Under Buhari Than Me—Jonathan
Source: Vanguard (Lagos), 21 November 2018
Author: Soni Daniel

> Quote 6: *Bribery and corruption are two of the prodigious factors that have derailed Nigeria from the path of progress.* They have also hindered the citizens from harnessing their potential. The year the country became independent (1960) was the same year independent countries in Africa started planning their developments. Many of the over 180 million population of Nigeria are still *dwindling in corruption.* According to Corruption Perception Index reported by

> Transparency International, Nigeria is currently the 154th corrupt nation out of 180 countries sampled. The principal index used for the ranking has to do with *how corrupt the public sectors of the countries ranked were perceived to be*. The index brought into the limelight the gravity of the damage done in the public sector, which is key to development.

Title: Bribery, corruption causing national backwardness
Source: The Punch, 10 January 2023
Author: Mukhtar Kobi

> Quote 7: A newly released global report on the interplay between democracy and corruption points to perhaps *Nigeria's biggest challenge*. In its September 2023 report, an NGO, the Open Society Foundations, traced the failings of democracy in Nigeria to pervasive corruption. This aligns with a global consensus that *corruption in Nigeria is endemic and hinders its development*. . . . Corruption manifests in every sphere of national life; most noticeably in elections, the judiciary, government agencies, procurement/contracts, payroll, and in extortion by security agents. For many, corruption is "*the single greatest obstacle* preventing Nigeria from achieving its enormous potential".

Title: Nigeria sinks deeper in corruption mire
Source: The Punch, 20 September 2023
Author: Anonymous
Theme: Describing and critiquing the CPI

> Quote 8: *Contrary to claims by the federal government* of significant gains in its anti-graft policy, new statistics by the Corruption Perception Index (CPI) 2017 has listed Nigeria as slipping downward in the international ranking. . . . The Corruption Perception Index *is believed to be* one of the most respected international measurement of corruption trends, established in 1995 as a composite indicator used to measure *perceptions of corruption* in the public sector in different countries around the world.

Title: Nigeria Slips 12% Downward In Global Corruption Perception Index, Group Alleges
Source: Daily Independent (Nigeria), 22 February 2018
Author: Innocent Oweh

> Quote 9: Presidential Advisory Committee Against Corruption (PACAC) chairman Prof Itse Sagay (SAN) yesterday described the latest Corruption Perception Index (CPI) released by Transparency International (TI) as *erroneous*. . . . He said: "Transparency International is making a *major error. It is confused between the actual level of corruption and the level of revelation of acts of*

> *corruption.* Corruption was worse than now. . . . Corruption is being dealt with. So, that's the error there." PACAC Executive Secretary Prof Bolaji Owasanoye said *the index relies on public opinion,* but that fighting corruption was an ongoing process. "The way the study is conducted is that public opinion is sought on perception", he said. According to him, *people may perceive corruption as being high because of low conviction rate.*

Title: Sagay: Transparency International report erroneous
Source: The Nation (Nigeria), 23 February 2018
Author: Anonymous

> Quote 10: President Muhammadu Buhari Friday hit back at Transparency International over its corruption perception index indicating that corruption was getting worse in Nigeria. Buhari, in a statement by his Senior Special Assistant on Media and Publicity Malam Garba Shehu, described as *misleading and unfair,* the ranking of 148 given Nigeria out if 180 countries surveyed. Shehu said President Buhari asked the anti-corruption watchdog *to focus on facts, not fiction.*

Title: Buhari hits back at Transparency International
Source: Weekly Trust, 24 February 2018
Author: Anonymous

> Quote 11: The statement reads: 'While TI's index only measures perceptions of corruption, their findings correspond substantially with the reality of impunity of perpetrators, as demonstrated by the low conviction rate, the authorities' slowness to adopt and implement critical reforms, appearance of selectivity in the anti-corruption fight, apparent complicity of key officials and cover-up, as well as unaddressed alleged corruption against several state governors. *The authorities should take the report seriously* and use it as an opportunity to raise their game in their efforts to rid our country of corruption and underdevelopment. The government should make no mistake about this: Nigerians know corruption when they see it. *Rather than simply criticising TI as publishing fiction or going after its board members,* the government should accept the fact that its oft-expressed commitment to fight corruption has not gone to plan. *The CPI may not be perfect, and in fact no index is.* The CPI may not show actual evidence of corruption in the country, but *perceptions are commonly a good indicator of the real level of corruption.* In any case, the *devastating effects of corruption* in virtually all sectors providing essential public services are too glaring for Nigerians to ignore.'

Title: TI's corruption index: A "wake-up call" for Buhari—SERAP
Source: Sunday Trust, 25 February 2018

Author: Anonymous

> Quote 12: Major reasons why the Buhari administration rejected the latest corruption perception index released by Transparency International emerged last night with the government, *claiming that the report was off the mark and did not capture its consistent fight against graft*. President Buhari On the basis of its sustained anti-corruption drive, which it claimed had resulted in blocking major corruption sources in the country, the Presidency promptly renounced the TI's report as a fiction geared towards embarrassing the government. . . . Hon. Reuben Yakubu Izeze, a member representing Ughelli South in the Delta State House of Assembly, stated that Transparency International is absolutely correct in its report on Nigeria. He said: "*The perception index is indeed an irrefutable truism*. The rebuttal by the Federal Government is *simply in her character of propaganda and falsehood*".

Title: The Truth, The Fallacies Of TI Report On Nigeria
Source: Daily Independent (Nigeria), 4 March 2018
Author: Anonymous
Theme: Factors causing/influencing corruption in Nigeria

> Quote 13: Having more accountants as a percentage of the workforce is correlated with better outcomes in Transparency International's Global Corruption Perception Index, according to IFAC's Nexus 3 study. Fraud and corruption thrive in the dark. Professional accountants operate in the public interest and a strong accounting profession, focused on implementing high-quality consistent global standards, helps to shine a light on financial management and makes it more difficult for corruption to occur.

Title: Nigeria can boost business confidence with accountability—Choudhury, IFAC boss
Source: The Sun (Nigeria), 29 January 2018
Author: Uche Atuma

> Quote 14: Highlighting Nigeria's case, NAS Capoon, Abiola Owoaje said based on global research by non-governmental organisation, Human Environmental Development Agenda HEDA, Nigeria lost $600 billion to corrupt government since her Independence; while she ranked 144th out of 180 nations on Transparency International Corruption Index of February 2019. 'Sadly, the *war against corruption in Nigeria has been viewed with suspicion by the citizenry because of the perception that it is selective, ineffective, and has therefore stumbled on without credibility*. The consensus opinion is that the battle against corruption must move beyond the tokenism of occasional recovery of funds and often theatrical convictions to the hard task of tackling the underlying structural defects,

institutional failure, mediocre political leadership and other sundry predisposing factors,' he said.

Title: Nigeria Yet to Feel Impact of Buhari's Anti-Corruption Fight—NAS
Source: Leadership (Abuja), 10 December 2020
Author: Chinelo Chikelu

Quote 15: Transparency International (2006) defines corruption as the abuse of entrusted power for private benefit. In the world around us, especially the country we live in, corruption is regarded as a given. Corruption is one of many intractable social pathologies ravaging Nigeria. British Prime Minister, David Cameron, without mincing words, described Nigeria as *"fantastically corrupt"* at the anti-corruption summit hosted in the UK on May 2016. Furthermore, the latest report on the 2021 Corruption Perception Index (CPI) by Transparency International rated the country as the second most corrupt country in West Africa. This makes it evident that *corruption is a social menace* that has eaten deep into the fabric of the Nigerian polity. . . . Today, Nigerians *applaud and tolerate ill-gotten wealth which in reality is money stolen from public coffers*. This is a pointer to the fact that *corruption is endemic* in Nigeria. It has *permeated into every facet of the society*; the family, the church and even the traditional systems are not left out of this contagious disease.

Title: Nigeria And The Corruption Virus
Source: News Chronicle (Nigeria), 6 February 2022
Author: Abdulrasheed Akere Abdulkareem and Ezinwanne Onwuka

Quote 16: In the 2022 Corruption Perception Index of Transparency International, Nigeria ranked 150 out of 180 countries; in 2021 it ranked 154, after slight improvements in 2015 and 2016. The 2022 Basel Anti-Money Laundering Index scored Nigeria 6.77 out of 10, finishing a poor 17th out of 128 countries. *It is neither for lack of intention, nor of laws, and policies*. . . . In the years ahead, analysts reviewing how the war floundered will cite Buhari's *customary inattentiveness, his lack of hands-on supervision, other leadership lapses, cronyism, and failure to rein in members of his inner circle and heads of key agencies, and the influence of the corrupt political elite among others*. All these and institutional weakness played their part. But perhaps the most decisive is this: Buhari did not realise that in a diverse, polarised polity like Nigeria, sectionalism in appointments and policies, nepotism and exclusionary practices are also *enablers of corruption* that were bound eventually to weaken and subvert the war. From the start, therefore, the war was doomed.

Title: The Buhari Years: How the anti-corruption war floundered
Source: The Punch, 5 May 2023

Author: Anonymous

> Quote 17: Prof. Angela Oyo-Ita, Deputy Vice Chancellor, (DVC) of the University of Calabar on Saturday said *corruption in Nigeria was more infectious than the dreaded and highly infectious COVID-19*. . . . She told the youth who were using the programme to commemorate the International Youth Day that they still had the power to choose not to be corrupt, to save the future of the nation. "Nigeria is 154th in the World Corruption Perception Index out of 180 nations with a 24 per cent rate of transparency, this *gives the nation a bad image*, we must change the trend. . . ." *The choice to be corrupt or not lies with us as a people* but if we must save the soul of this nation for the future generations, we need to choose the latter", she said. . . ." As a church, we must promote the three 's' model which is to *shun, speak and stand against corruption* in our politics, educational system and even in the church to save our nation.

Title: Corruption, more infectious than COVID-19, says Unical DVC
Source: The Sun (Nigeria), 13 August 2022
Author: Ajiri Daniels

The first theme highlighted above is titled "Extent of corruption and failure of governance" and perhaps captures the flavour of most of the stories in the Nigerian newspapers. Many of the stories use the CPI as the foundational, and external, evidence for showing how corruption is rampant in the country and how this is linked to a failure of governance. This is by no means a new insight, and the academic literature has long highlighted and explored the prevalence of corruption in Nigeria and indeed throughout Africa and its impact on sustainable development (Akanle and Shittu, 2022; Hope, 2024). As we would perhaps expect, the incumbent government gets much of the blame in these newspaper stories, and this is usually levelled at them by their opponents, but the stories capture blame at all levels in society including within the private sector. The stories often bemoan the failure of interventions designed to root out corruption, with one of the stories dismissing the "current anti-corruption fight as mere media show", while others point to politicians seeking to downplay the problem—as one of the quotes says

> Some people may be misled with smoke and mirrors but the TI Corruption Perception Index relies on unsentimental facts and figures.

Given that the CPI is based on people's perceptions, it would seem something of a stretch to regard them as "unsentimental facts and figures", but the indicator is clearly seen as an antidote to the kind of "spinning" that often appears throughout these quotes and is an example of the point made by Lowrey and Hou (2021) of the use of reports published by respected bodies to add legitimacy to stories. Even so, the pervading sense throughout these stories is that corruption is a major issue in Nigeria and is having dire consequences for the country. Recognising that there is a problem is always a necessary first step to finding a solution, and the CPI has gravitas given its publication by a highly respected and international NGO.

The second theme of quotes is called "Describing and critiquing the CPI". Here, the quotes are representative of a group of stories that describe what the CPI is and how it is arrived at, and in some cases, that background is used as the basis for critiquing its messaging. For example, a number of these stories appear to revolve around a critique from President Buhari and his staff that the CPI does not really capture efforts being made by them to control corruption. Indeed, there is a statement here that

> President Buhari asked the anti-corruption watchdog [i.e. Transparency International] to focus on facts, not fiction.

But there appears to be a mismatch between what the CPI does and what the politicians think it should do. The CPI is based on a snapshot of people's perceptions as to the level of corruption in the country and is not designed to capture what people feel about plans to address it. Hence it would seem odd for some to critique the CPI for not doing what it was not designed to do. But it can nonetheless be a useful deflection for politicians to claim that the CPI is somehow not measuring the "right thing" (i.e., what they see as the "facts") and is instead engaged in measuring "fiction". A further intriguing comment in the quotes is that "people may perceive corruption as being high because of low conviction rate [for corruption]"; the assumption here appears to be that the low rate of conviction of those accused of corruption leads to perceptions of a higher rate of corruption and thus a worse CPI. Basically, if there is a low conviction rate then people are led to the conclusion that corruption must be high as there is no disincentive for being corrupt. Following the same logic, higher conviction rates of corruption would presumably help generate the perception that corruption was somehow lower as those who engage in corruption would be dissuaded by the high conviction rates. The logic behind all of this seems somewhat convoluted and surely there are simpler explanations for high levels of perceived corruption. Maybe the level of corruption is so high that it encompasses law enforcement agencies and becomes tolerated and ignored, and this would explain how a low conviction rate can occur alongside high levels of perceived corruption. But there are other stories which defend the CPI from these attacks by politicians and point out that while the indicator is not perfect and is acknowledged to be based on perception, it nonetheless provides valuable insights. For example, in one of the stories we have the following strong statement "The perception index is indeed an irrefutable truism. The rebuttal by the Federal Government is simply in her character of propaganda and falsehood".

The third theme is titled "Factors causing/influencing corruption in Nigeria" and while the stories often begin with a statement of the extent and importance of corruption in Nigeria, with the CPI provided as evidence to support that claim, the story moves on to try and tease out some of the main causes of corruption beyond blaming politicians for failing to control it. Only a few of the potential causes are captured in these quotes. One of them suggests that what Nigeria needs is more accountants who can forensically analyse monetary flows and thus detect corruption, although the article does not address the question as to what happens if the accountants become corrupt! Second, there is the argument that people in Nigeria regard the various "Wars against corruption" waded by successive government as being ineffective and something stronger is required. Hence, the "battle against corruption must move beyond the tokenism of occasional recovery of funds and often theatrical convictions" to addressing many of the deep-seated issues at the heart of the problem, although the latter does seem to be a rather lengthy and challenging list. Above all is the recognition that corruption is endemic in Nigeria and has become widely accepted and tolerated. Indeed, as one of the quotes notes, Nigerians often "applaud and tolerate ill-gotten wealth which in reality is money stolen from public coffers", and what is needed is to "shun, speak and stand against corruption" throughout society. The assumption running through the endemic argument is that the occasional trial for corruption is just "tokenism" and "theatrical", and what is needed is something far deeper and pervasive.

The second group of quotes is from newspapers between 2020 and 2023 that mention the CPI in relation to Australia. The number of articles in this group is much smaller than for Nigeria, and many of the newspapers carrying the story are not necessarily based in Australia. Hence, unlike the sample of quotes shown above for Nigeria which have all been taken from newspapers based in that country, many of the quotes in the sample below have come from newspapers based in a wide range of countries, although most of them are from Asia. As before, the emphases (bold text and underline) are all mine.

Theme: Regional comparisons of corruption

"New Zealand has regained its spot as the *worlds least corrupt country* in a major annual study, but Justice Minister Andrew Little admits there's still work to be done and a more powerful corruption agency may be needed. Transparency Internationals 2019 Corruption Perception Index has given New Zealand a mark of 87 out of 100, tied first with Denmark out of 180 countries. New Zealand has been a high performer in the annual study—the most closely watched of its kind—vying for first place the past eight years and also topping the chart in 2017. . . . The news was less positive across the Tasman, with *Australia among the 21 countries that have declined the most* in the past year, to a score of 77".

Title: New Zealand ranked worlds least corrupt country
Source: The New Zealand Herald, 24 January 2020
Author: Anonymous

"Malaysia has jumped 10 spots from last year to 51st place on Transparency International's (TI) 2019 Corruption Perceptions Index (CPI), as a result of the government's actions against financial scandals at Malaysia Development Bhd, the Federal Land Development Authority and Lembaga Tabung Haji. . . . On the *regional front*, the Asia-Pacific region—which includes *top performers New Zealand, Singapore and Australia*—scored an average 45 points, coming in second behind Western Europe and the European Union which averaged 66 points".

Title: Malaysia jumps 10 spots in TI's Corruption Perception Index
Source: The Malaysian Reserve, 24 January 2020
Author: Alifah Zainuddin

"India's ranking on the Corruption Perception Index-2020 slipped from 80 to 86 even as its score decreased only by one point to 40 from 41 in 2019. . . . The average score in the Asia-Pacific region, which encapsulates 31 countries, is 45. India's score of 40 is *below both the global average and the Asia-Pacific average*. India's score is also lower than that of China, which scored 42 with a rank of 78. However, Pakistan fared poorly with a score of 31 and a rank of 124. In the Asia Pacific region, New Zealand was the top performer. Other *topscoring nations* were Singapore (85), Australia (77) and Hong Kong (77)".

Title: India slips six spots to 86th rank on corruption index
Source: Mirror Publications, 30 January 2021
Author: Anonymous

"The 2021 Corruption Perception Index by Transparency International *places India 85th on a list of 180 countries*, one position above last year. Not much has changed about how corrupt India is perceived to be in a decade. . . . Among larger economies, only Italy, China, Austria, UK and Argentina have improved their scores. The report flagged the *backslide of "advanced economies" like Australia* (from 85 to 73), Canada (84 to 74) and the US (73 to 67) over the past 10 years".

Title: India stuck on corruption score, US slides
Source: The Times of India (TOI), 26 January 2022
Author: Anonymous

"While it improved a notch, the Philippines has retained its score and *remains among the 'significant decliners' among Asia-Pacific countries* included in the latest Corruption Perception Index (CPI) of Berlin-based Transparency International. The 2022 CPI released yesterday showed that the Philippines scored 33 out of 100, below the global average of 43. It ranked 116th out of 180, up from 117 in 2021. . . . Other Asia-Pacific countries included in the latest *report's 'significant decliners' list* are Malaysia (47, down six points since 2019), Pakistan (27, down six points since 2018), *Australia (75, down four points since 2016)*, Mongolia (33, down four points since 2018) and Sri Lanka (36, down four points since 2012)".

Title: "Significant decliner" Philippine ranks 116th in global corruption index
Source: The Philippine Star, 1 February 2023
Author: Anonymous

Theme: Causes and control of corruption

"Anti-graft experts have intensified their calls for a robust federal integrity watchdog amid the *sports rorts affair* and a sustained slide in Australia's position in global corruption rankings. A report by Transparency International shows Australia's Corruption Perception Index reading was 77 points in 2019, *an eight point decline since 2012*, driven down by concerns around *opaque political financing and the undue influence of corporate interests on political decision making*. Australia still ranks highly, 12th out of 180 countries, reflecting *relatively low levels of public sector corruption*, but it has been highlighted by Transparency International as one of three 'decliners' (alongside Canada and Nicaragua) and is no longer in the top echelon of countries considered to be clean, such as Denmark, New Zealand, Singapore and Sweden".

Title: Sports rort fuels national integrity push
Source: Canberra Times (Australia), 25 January 2020
Author: Adrian Rollins

"Momentum for an *effective federal anti-corruption body* is an "unstoppable force", the government has been warned, as transparency advocates criticise its proposal as "worse than no commission at all". . . . But Transparency Australia chief executive Serena Lillywhite said a groundswell of support for a powerful watchdog, including from independent candidates and crossbenchers, made its implementation a matter of time. . . . Australia has slipped in The Transparency International Corruption Perception Index over the past eight years, dropping out of the top ten for the first time since 2014. Although Australia's standing was "pretty decent" in global terms, Ms Lillywhite said its negative trajectory was "the critical issue". But she warned the government's model would simply entrench that distrust by establishing different rules for politicians and members of the public. "The introduction of a poorly designed commission is worse than having no commission at all", she said".

Title: Unstoppable force: Government warned over federal ICAC
Source: Canberra Times (Australia), 2 December 2021
Author: Finn McHugh

Theme: Corruption, tourism and the threat to Australian businesses trading internationally

"Corruption is often endemic in fast-growing economies where the greatest opportunities exist. Two thirds of the countries on Transparency International's corruption perception index score below 50 out of 100. China scores 41. Like the US, UK and other Western countries, *Australia regards foreign bribery as a domestic crime*. The criminal code imposes 10-year jail terms and $21 million corporate fines for *bribery of foreign officials*, for good reason. Payments for government contracts undermine democracies and perpetuate economic inequality. They can transfer wealth from shareholders to corrupt managers. . . . For regulators and investigators, prosecuting foreign bribery can be a nightmare. *Obtaining evidence from overseas is often slow and difficult and requires expertise in business, finance and international trade*. The cases can cause *diplomatic tensions*. . . . Given the challenge of winning contracts overseas without paying bribes, it is not surprising that many Australian companies prefer the predictability of home and New Zealand. Given the *difficulty proving and prosecuting foreign corruption*, and the perception that they are a normal part of business in some countries, it is not surprising that some Australian executives take the chance too".

Title: The challenge of not being corrupt
Source: Australian Financial Review, 13 January 2021
Author: Aaron Patrick

"Australia is on the verge of getting a new Pacific neighbour, but experts advise caution if you're planning an island holiday there. Bougainville voted for independence from Papua New Guinea in a referendum landslide in 2019, starting a process that will begin in 2023 which its citizens hope will lead to full independence by 2027. . . Plus, it still has *many factors that keep its parent nation PNG off the beaten track when it comes to tourism*—especially a perilous reputation for personal safety. Much of this is due to crime and violence between locals (especially domestic violence and violence towards women) and corruption. Transparency International's 2016 Corruption Perception Index *ranked PNG 142nd from 180 countries*".

Title: Australia is getting a new neighbour! The island of Bougainville is gaining its independence from Papua New Guinea—but you might not want to book a holiday there just yet
Source: MailOnline, 24 November 2021
Author: Peter Vincent

The first group of quotes have been placed within the theme of "Regional comparisons of corruption", and this would seem to be the most common theme across the articles. They typically involve little more than the CPI (and ranking) of Australia being provided alongside those of other countries, perhaps with some comment about how the countries compare. In each of these stories, there is little discussion of corruption in Australia per se, just a brief mention as to whether it is performing well or badly in the rankings relative to the other countries. The CPI for Australia is thus used as a sort of "waypoint" for comparison, and this tendency is especially so for Asian countries as can be seen in the quotes that span New Zealand, Malaysia, India and the Philippines, with accompanying comments such as the "backslide of advanced economies like Australia" in the CPI rankings. The period covered by these articles was not a good one for Australia in the CPI rankings, although as one story noted "Australia's standing was 'pretty decent' in global terms".

The second theme titled "Causes and control of corruption" comprises newspaper articles that are more focussed on Australia rather than the country simply providing a "waypoint" for comparison in the CPI rankings. One of these stories makes the case that Australia's slide in the CPI rankings was largely due to "opaque political financing and the undue influence of corporate interests on political decision making" and specifically cites the "Sports Rorts" affair which happened between 2018 and 2019, with a report published in 2020, as an example. This is also called the "McKenzie scandal" after Senator Bridget McKenzie who was at the time the Minister for Sport and was based on allegations that the government was using funds allocated to sport to support their political party in marginal constituencies; in other words, the governing party was accused of misappropriating funds to garner political advantage (Ng, 2023). Another of the quotes refers to ongoing debates at the time in Australia for the establishment of an "effective federal anti-corruption body" and criticisms of what the government had in mind for this. The final theme titled "Corruption, tourism and the threat to Australian businesses trading internationally" has two quotes from stories that have occasionally surfaced in Australian newspapers. The first quote is from a story that focusses on the need to prosecute Australian companies if they are alleged to have been involved in corrupt business practices in other countries. The moral here is that just because those practices may have been alleged to take place in other jurisdictions, it is Australian law that holds even if there can be challenges when it comes to getting the necessary evidence to support a prosecution. The final quote is linked to tourism, and here the CPI is employed to highlight the dangers involved of being a tourist in Papua New Guinea; personal safety may be an issue for the traveller but so too could corruption.

These two sets of quotes for Nigeria and Australia present us with two very different patterns. For Nigeria, the stories are inward-looking and largely centre on the extent of corruption in the country, its impacts and control. As the CPI is published by a major international NGO, it acts as a focal point for the stories, a sort of respected yardstick which shows the state of play in the country. Given the nature of the CPI, it is perhaps unsurprising to see some stories reporting how politicians and those who work for them trying to downplay its significance, and part of this "attack" is rooted on the perception-based nature of the indicator. Hence, some in power have claimed it is basically the wrong type of indictor; what is really needed is an indicator that measures what government is doing about corruption (i.e., the response) rather than an indicator based on perceptions of the extent of the problem (i.e., the state).

By way of contrast, for Australia, the bulk of the reporting of the CPI is more outward-looking, minimalist and comparative; the stories mostly use Australia as a "waypoint" for making comparisons, although over this period, Australia was sliding in the CPI rankings due to the "Sports Rorts" affair and concerns expressed by some over initiatives to control corruption. There is little in the way of discussion focussed on Australia in many of these "comparator" type stories, and instead, the main emphasis is typically on another country, and Australia is included for comparison. The stories that do focus on Australia tend to use the CPI as an evidence-base for the challenge of corruption, before moving on to discuss its control.

5.6 A web of influences: Whose reality really counts?

It is not unreasonable to suppose that the type and degree of influence of both perception-based and non-perception-based indicators will be related to the degree of exposure that these tools have attracted within groups such as the media. Thus, it seems reasonable to assume that an indicator that attracts very little attention would have little chance of having any influence on the public, politicians and others. But, of course, this influence arising from media reporting of indicators is by no means a one-way street, and the media itself is influenced by other actors. The media is not only a window but also a mirror. Therefore, the graphs shown in Figures 5.1 and 5.2 can perhaps also be seen as capturing a degree of influence; the less reporting there is of an indicator; in this case, the number of articles that mention it, then the lower may be its influence. But this picture is a simplistic one as it is based on reporting in just one form of traditional media, i.e., newspapers. There are, of course, other forms of media, and indeed, social media platforms have become an important source of information and news for many. Hence, Figures 5.1 and 5.2 only represent one perspective—one window—on indicator reporting, and it is entirely conceivable that those indicators that do badly in terms of reporting in newspapers may do far better in other forms of media. This is certainly an area that is ripe for further research.

Given the points made above, one cannot help but wonder whether perception-based indicators have an advantage over their "non-perception"-based cousins when it comes, for example, to reporting via social media? After all, perception-based indicators are often designed to capture hard-to-reach issues such as crime and happiness that are important and of concern to many and have a feel of direct connection to people given that the perceptions are those of broader groups in society and not only experts. Hence, it can perhaps be hypothesised that perception-based indicators may have a more "personal" feel. This rationale is related to the hypothesis set out in Chapter 4 for perception-based indicators emerging out of a participatory process and is admittedly something of a stretch, and there is no evidence available that I am aware of to back it up. People care about crime and happiness, but they also care about income and well-being, and while perception-based indicators are founded on the views of people rather than the kind of data collection behind GDP and HDI, does that make them resonate better with a wider public? Perhaps the difference here is not so much about the perception versus non-perception-based nature of the indicators, but that perception-based indicators are trying to shine a light on areas that are challenging to reach using other means of data collection. One possible way of exploring this issue is to look at reporting of indicators within social media. While the newspaper reporting covered in the previous sections were written, at least mostly, by professional journalists, social media platforms are open to anyone to use. Do the patterns of social media reporting of perception-based indicators differ from newspaper reporting?

It is possible to do the same kind of analysis as that above for newspaper, but this time for social media reporting of indicators using what used to be called Twitter but is now called "X". There are numerous social media platforms that could have been chosen for this foray, but "Twitter" (X) has its advantages:

> The microblogging platform Twitter stands as the most appealing context in which to analyze the processes of interaction and the circulation of scientific knowledge, as it is used by a wide-ranging number of academic and non-academic users. These users discuss issues that span from local to global topics and from professional to personal interests.
>
> *(Dıaz-Faes et al., 2019; pages 2–3)*

Figure 5.4 shows the number of tweets made each year between 2006 and 2020 that mention the HDI (a non-perception-based indicator) and the CPI (a perception-based indicator) at least once. For context, the number of tweets is shown alongside the number of newspaper articles, with the latter being the same data used for 2006–2020 in Figures 5.1a and 5.2a for the CPI and HDI, respectively. The period 2006–2020 was chosen to span the very first tweet sent (2006) to the time of the onset of the COVID pandemic in 2020. Many of the country-wide lockdowns across the globe began in March 2020, and during the lockdowns, Twitter showed major growth in members and volume. In 2022, it was acquired by Elon Musk and rebranded as "X" in July 2023. Hence, the lockdowns in many countries were something of a turbulent period for social media. I will also use the older name of "Twitter" which applied during 2006–2020, and a "tweet" is an individual posting on Twitter. It does also need to be noted, of course, that a degree of interaction in reporting of indicators via all social media platforms, not only Twitter, and reporting in the traditional media is to be expected. People writing posts in social media may well be picking up on reports they have seen in the traditional media, and such an influence may potentially be vice versa.

The counts span the period 2006–2020; the first tweet was sent in 2006. Correlation coefficients (r) for the number of newspaper articles and tweets each year are 0.822 (df=13; $P<0.001$) for the HDI and 0.554 for the CPI (df=13; $P<0.05$).

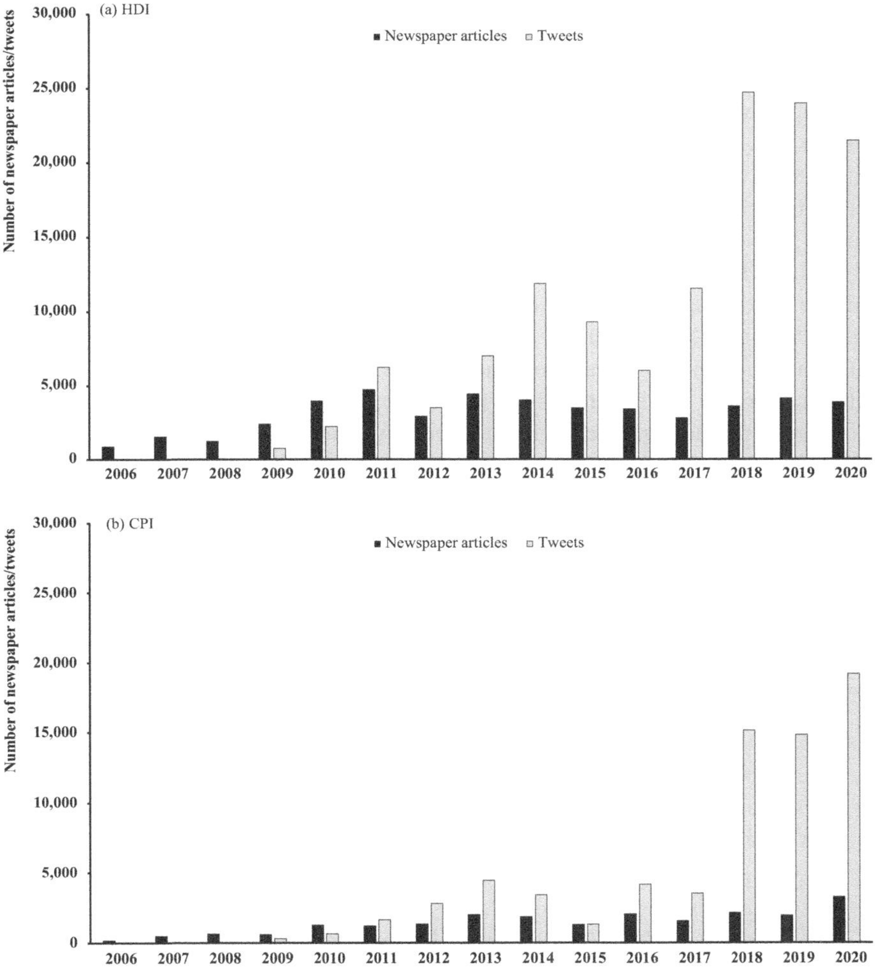

FIGURE 5.4 Number of tweets that mention the HDI (a) and CPI (b) at least once alongside the corresponding number of newspaper articles

Taking the period 2006–2020 (15 years), the number of "pieces" (newspaper articles and tweets) that mention the indicator at least once from 2011 onwards is higher for tweets (grey bars) than for newspaper articles (black bars) and is unsurprising given that the number of people writing tweets is likely to be much greater than the number of journalists. But the surge in the number of tweets from 2018 to 2020 for both indicators is intriguing, especially as an equivalent surge was not seen with the newspaper reporting. Nonetheless, there is evidence to suggest that for both the HDI and CPI, a correlation exists between the number of newspaper articles and tweets. For the CPI, the correlation between these counts is 0.554 (statistically significant at $P<0.05$), while for the HDI, it is 0.822 (statistically significant at $P<0.001$). We can go a little further and check whether the form of the apparent relationship is the same for both indicators, and the analysis is set out in Box 5.1. The results suggest that the number of tweets has a stronger link to the number of newspaper articles for the HDI than it does for the CPI. Given that the HDI (first published in 1990) is only some 5 years older than the CPI, it is interesting to speculate why the link between tweets and newspaper articles is stronger for the HDI. Maybe it comes down to an assumption that the HDI measures "quality of life" and "wellbeing" and thus of much potential interest to a wider range of people. Ironically, the HDI is not a measure of either, as its curators continuously emphasise, but it is not hard to imagine such an assumption, even if false, would attract attention and be circulated via social media platforms. By way of comparison, the CPI is perhaps seen as having a narrower focus and thus attracts less attention in social media.

BOX 5.1 EXPLORING DIFFERENCES BETWEEN THE HDI AND CPI IN TERMS OF HOW TWEET COUNTS/YEAR (DEPENDENT VARIABLE) RELATE TO NEWSPAPER ARTICLE COUNTS/YEAR (INDEPENDENT VARIABLE).

STEP 1: Using data summarised in Figure 5.4 for tweet and newspaper article counts per annum for the HDI and CPI, the first step is to see whether the tweet counts are related in a similar way to newspaper article counts. But to do this, it is necessary to take the logarithm (base e) of the raw data where:

$$\text{Transformed value} = \text{Logarithm}(\text{raw data value} + 1)$$

The one is added because some of the tweet counts are zero. The transformation applied has the effect of contracting the data and making the relationship between the variable more linear and hence simpler.

STEP 2: The transformed variable can now be plotted against each other and learn regression models fitted. The result is shown in Figure 5.5.

The two dotted lines are "least square" regressions (linear model) for the HDI and CPI. Even based on visual appearance, it does look like there is a strong relationship between tweet and newspaper counts per year. In fact, for these transformed data, the correlation coefficients (r) are the same for the two indicators at 0.94 ($P<0.001$). But the two "best fit" models seem to have different slopes and indeed intercepts on the x-axis—they do not look the same. But how can we test this hypothesis of the models being different?

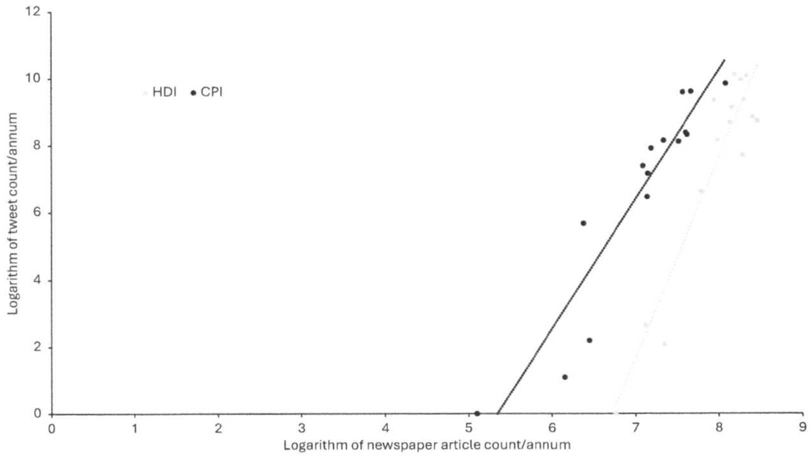

FIGURE 5.5 Logarithm of tweet counts per annum as a function of the logarithm of newspaper counts per annum, for two indicators (HDI and CPI)

STEP 3: One way of testing for a difference in fitted regression models is to use ANOVA to compare them. This approach uses the same transformed data derived in step 1 but adds another "dummy" variable in the analysis for indicator—in this case, 0 for the HDI and 1 for the CPI. It does not matter which way around you do this—it could be 1 for the HDI and 0 for the CPI. The inclusion of a dummy variable for indicator allows us to test for an interaction between indicator and newspaper article counts.

However, before doing the analysis including an interaction term, it is necessary to centre the tweet and newspaper count variables, and this is achieved by subtracting the mean from each variable before the analysis is undertaken. Centring the data in this way helps address a potential problem of interaction terms having high "variance inflation" as described in Chapter 3, a sort of "double counting" of influence on the regression which arises from variables that may be highly correlated. Centring results in a new variable with a mean of zero, and while this technique has its advantages, it does not always work when it comes to reducing collinearity.

Plotting the centred data generates Figure 5.6, which may look odd at first glance given that some of the values for the CPI have now become negative, but the overall appearance of the graph and best-fit regression lines is similar to Figure 5.5 using un-centred data.

STEP 4: Once the dummy variable for indicator has been included and the tweet and newspaper article data have been centred, the data can be analysed—with the inclusion of an interaction term—and the results of the ANOVA

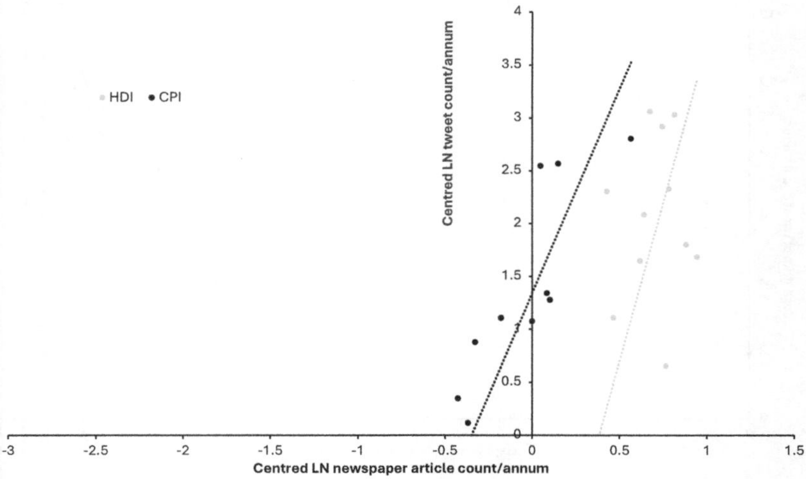

FIGURE 5.6 Logarithm of tweet counts per annum as a function of the logarithm of newspaper counts per annum for two indicators (HDI and CPI), but the tweet and article count data have been centred

are shown in Table 5.1(a). The regression coefficients for the "best fit" model between tweet and article counts that account for the two different indicators and their interaction are shown in Table 5.1(b). Note that because of the "centring" in step 3, the VIFs are relatively low for all the independent terms—well below the 5% "point of concern" mentioned in Chapter 3 for VIFs—so that is encouraging. Without centring the data, the VIFs for some terms would be very high—nearly 200%!

TABLE 5.1 Results of an ANOVA to compare two regression models fitted to the relationship between tweet and newspaper article counts for the HDI and CPI.

(a) ANOVA Table

Source of variation	*Degrees of freedom*	*Adjusted sum of squares*	*Adjusted mean square*	*F-value and significance*
LN Newspaper article counts	1	239.33	239.33	186.77***
Indicator (dummy variable)	1	61.86	61.86	48.27***
Interaction: LN Newspaper article counts X Indicator	1	11.18	11.18	8.73**
Error	26	33.32	1.28	
Total	29	285.55		

Adjusted $R^2 = 83.81\%$ (s=1.132)

(b) Regression coefficients

Model term	*Coefficient (Standard Error)*	*t-value and significance*	*VIF*
Constant	−0.477 (0.262)	−1.82 ns	
LN Newspaper article count	4.927 (0.361)	13.67 ***	1.79
Indicator (dummy variable)	−1.822 (0.262)	−6.95***	1.61
Interaction: LN Newspaper article count X Indicator	1.065 (0.361)	2.95**	1.18

Note: The analysis was performed on the "centred" article and tweet count data described under step 3.

STEP 5. Interpretation. The relationship between the "centred" tweet and newspaper article counts is statistically significant at $P<0.001$, and indicator also has an influence which is significant. Both conclusions may be anticipated from the graphs above; tweet count does seem to increase with newspaper article count, and tweet counts (and newspaper article counts) do seem to be higher for the HDI than the CPI. But the key term we are interested in is the interaction between newspaper article count and indicator, and this is also statistically significant at $P<0.01$ (less than 1% probability). This tells us that there is indeed an influence of indicator on the relationship between tweet and newspaper article counts. But what is the nature of the influence coming from the indicator? This can be answered by looking at the best fit regressions for each of the two indicators:

$$\text{HDI}: \text{LN centred Tweet counts / annum} = -2.299 + 5.992\text{LN centred Newspaper counts / annum}$$

$$\text{CPI}: \text{LN centred Tweet counts / annum} = 1.345 + 3.862\text{LN centred Newspaper counts / annum}$$

Based on the two "slope" terms (5.992 for the HDI and 3.862 for the CPI), it seems that newspaper counts for the HDI are having a stronger influence on tweet counts than is the corresponding case for the CPI. The question remains as to why that should be?

Finally, a word of warning which echoes points already made in Chapter 3 when it comes to any cause–effect assumptions behind correlation and regression. The assumption in the above analysis is that the count of tweets is being influenced by newspaper coverage of the indicators (i.e., count of newspaper articles). But it is also possible that coverage of the indicators in tweets influences newspaper reporting. In some ways, this does not matter—the key point being made with the analysis is that tweets and newspaper coverage of the indicators are related but seems to be more intense for the HDI relative to the CPI.

5.7 Summary

Indicators are designed to attract attention to an issue as a crucial first step to someone using them to bring about what the indicator curators regard as a desired change. This may seem like a somewhat basic statement, and indeed, it is, but it is after all the raison d'etre of all indicators; they have no value whatsoever unless they are communicated and then "used". Indicators are human constructs designed to help bring about a desired change. They are the "pop stars" of the data world; people must see them and like what they see as a prelude to someone "buying" into the message(s) they are intended to convey. Without that attention and influence, any indicator is useless. But assessing the influence of any indicator on the decision-making process is complex, not least because there are so many factors at play, many of which will interact, and because the influence of indicators may be subtle and occur over time. Indeed, it may well be the case that those who are influenced by an indicator may not even realise they have been; the influence may be quite subtle and may even be unconscious. This dimension of the indicator "multiverse" has arguably received less attention than it should. However, there has been research on the reporting of indicators by the media, which is a form of use as well as communication, and this can reasonably be assumed to at least be a starting point in terms of investigating influence. The evidence to date is limited but does suggest that indicators differ in terms of the extent to which they are reported by newspapers (as one form of traditional media), and the ways in which they are used to bolster claims in stories. But the research to date has just scratched the surface and much more needs to be done, especially in terms of the use of indicators in social media and how that may compare to the traditional media.

To date, there appears to be no evidence that perception-based indicators fare any better in terms of media reporting or indeed influence on decision-makers than do those indicators not based on perception. The counts of newspaper stories that make use of the perception and non-perception-based indicators do not suggest an emphasis on the former. If anything, the counts suggest that the GDP, an economic indicator, is used far more in newspaper stories than any of the other indicators. Of all the perception-based indicators, the only one that seemed to have significant traction in newspaper reporting was the CPI. Many of the perception-based indicators listed in Table 3.1 are hardly mentioned in this medium.

The difference in reporting of the CPI between Nigeria and Australia is intriguing. With Nigeria, which tends to rank badly in the CPI league tables, there are more newspaper stories that make use of the CPI compared to Australia. The publishing of the CPI by a respected international body, and an NGO at that, is often mentioned in these stories, and the indicator appears to spark regular debates in the media about the causes of corruption and what the country can do about it. But there are also those who would try and dismiss the messenger, and some of the stories illustrate one of the potential disadvantages of perception-based indicators. On the one hand, there are those in government who say that the basis of the CPI on gauging perception of corruption misses the point as it is actions to control corruption that really matter. The argument being employed here is that the CPI is thus based on "fiction" (i.e., perception of the level of corruption), while attempts by the government to do something about corruption are seen by government as "facts". Another argument used by critics is that they regard the perception of corruption having an inverse relationship to action; lower rates of prosecution of corruption cases may be perceived by the public as suggesting that the rate of corruption is higher than it is as less prosecution means that there is less disincentive for officials to engage in corruption. Hence, people's perceptions of the extent of corruption are being unduly heightened by fewer cases going through the courts; in effect, the public are drawing the wrong conclusion. This explanation seems somewhat convoluted, and there are others that would seem more straightforward. For example, perhaps lower numbers of successful prosecutions may be caused by a toleration or indifference towards corruption shown by enforcement agencies, and maybe those agencies also have high levels of corruption. This explanation would equally create a combination of low levels of prosecution alongside high levels of perceived corruption.

The reporting of the CPI linked to Australia is more benign and perhaps less intense than it is for Nigeria and given that Australia tends to rank towards the top of the CPI "league table" this is understandable. Australia is more of a "good news" story when it comes to corruption than is Nigeria, and there is well-reported evidence that the media tends to favour the reporting of bad news (Haskins and Miller, 1984; Soroka, 2006; Garz, 2014; Kätsyri et al., 2016; Kollmeyer, 2004). For the most part, the reporting of the CPI for Australia is one of the comparison with other countries in the region. Nonetheless, there are newspaper articles which have used the CPI as part of coverage of corruption cases in Australia; either directly, as with the "Sports Rorts" affair but also indirectly in relation to activities of Australian companies operating overseas. This reinforces a point made repeatedly by TI that there is no such thing as a country entirely free of corruption.

Finally, the chapter mentions the results of some research designed to explore whether reporting of indicators in the social media is correlated to reporting in the traditional media and by and large it is. The pattern was noted for both the HDI and the CPI and suggests there may be a rich interdependency, with all these forms of media feeding from each other. If anything, the relationship between the number of tweets and newspaper articles for the HDI is stronger than it is for the CPI. But based on these limited results, whether an indicator is perception-based or not makes little, if any, difference in the undoubtedly complex interplay between reporting within the traditional and social media.

5.8 Food for thought

1. What are the different types of use of an indicator, and which groups would be the most likely to have those uses?
2. What are the key features of any indicator that would make it attractive to the media?

3. How do you think the "nudge" effect of indicators could be measured in practice and on what groups would you focus?
4. How can we compare the influence of perception and non-perception-based indicators?

References

Aday S (2010). Chasing the Bad News: An Analysis of 2005 Iraq and Afghanistan War Coverage on NBC and Fox News Channel. *Journal of Communication* 60, 144–164

Akanle O and Shittu OS (2022). The Unending Development Question of Nigeria. *The European Journal of Development Research* 34, 321–342

Barabas J and Jerit J (2009). Estimating the Causal Effects of Media Coverage on Policy-Specific Knowledge. *American Journal of Political Science* 53(1), 73–89

Cairney P and Oliver K (2017). Evidence-Based Policymaking Is Not Like Evidence-Based Medicine, So How Far Should You Go to Bridge the Divide Between Evidence and Policy? *Health Research Policy and Systems* 15, 35

Dıaz-Faes AA, Bowman TD and Costas R (2019) Towards a Second Generation of 'Social Media Metrics': Characterizing Twitter Communities of Attention Around Science. *PLoS ONE* 14(5), e0216408

Eckerberg K and Mineur E (2003) The Use of Local Sustainability Indicators: Case Studies in Two Swedish Municipalities. *Local Environment* 8(6), 591–614

Gans H (1980). *Deciding What's News*. New York: Vintage Book

Garz M (2014). Good News and Bad News: Evidence of Media Bias in Unemployment Reports. *Public Choice* 161, 499–515

Gudmundsson H (2003). Making Concepts Matter: Sustainable Mobility and Indicator Systems in Transport Policy. *International Social Science Journal* 55(176), 199–217

Hallsworth M, Parker S and Rutter J (2011). *Policy Making in the Real World. Evidence and Analysis*. Institute for Government, London

Haskins JB and Miller MM (1984). The Effects of Bad News and Good News on a Newspaper's Image. *Journalism Quarterly* 61(1), 3–13, 65

Hezri AA (2004). Sustainability Indicator System and Policy Processes in Malaysia: A Framework for Utilisation and Learning. *Journal of Environmental Management* 73, 357–371

Hezri AA (2005). Utilisation of Sustainability Indicators and Impact Through Policy Learning in the Malaysian Policy Processes. *Journal of Environmental Assessment Policy and Management* 7(4), 575–595

Hezri AA and Dovers SR (2006). Sustainability Indicators, Policy and Governance: Issues for Ecological Economics. *Ecological Economics* 60, 86–99

Hezri AA and Hasan MN (2004). Management Framework for Sustainable Development Indicators in the State of Selangor, Malaysia. *Ecological Indicators* 4, 287–304

Hope, KR Sr (2024). The Corruption and Sustainable Development Nexus in Africa: A Contemporary Review and Analysis. *Journal of Financial Crime* 31(2), 331–346

Huston AC (2008). From Research to Policy and Back. *Child Development* 79(1), 1–12

Kätsyri J, Kinnunen T, Kusumoto K, Oittinen P and Ravaja N (2016) Negativity Bias in Media Multitasking: The Effects of Negative Social Media Messages on Attention to Television News Broadcasts. *PLoS ONE* 11(5), e0153712

Kollmeyer CJ (2004). Corporate Interests: How the News Media Portray the Economy. *Social Problems* 51(3), 432–452

Lowrey W and Hou J (2021). All Forest, No Trees? Data Journalism and the Construction of Abstract Categories. *Journalism* 22(1), 35–51

Lyytimäki J, Tapio P, Varho V and Söderman T (2013) The Use, Non-Use and Misuse of Indicators in Sustainability Assessment and Communication. *International Journal of Sustainable Development & World Ecology* 20(5), 385–393

McCluskey JJ, Kalaitzandonakes N and Swinnen J (2016). Media Coverage, Public Perceptions, and Consumer Behavior: Insights from New Food Technologies. *Annual Review of Resource Economics* 8, 467–486

Morse S (2019). *The Rise and Rise of Indicators: Their History and Geography*. Routledge, London

Morse S (2023a). *Ranking Nations: The Value of Indicators and Indices*. Edward Elgar, Cheltenham, Gloucestershire

Morse S (2023b). Quality of Life, Well-Being and the Human Development Index: A Media Narrative for the Developed World? *Social Indicators Research* 170, 1035–1058

Neylan, J (2008). Social Policy and the Authority of Evidence. *The Australian Journal of Public Administration* 67(1), 12–19

Ng Y-F (2023). Regulating the Rorts: The Legal Governance of Grants Programs in Australia. *Federal Law Review* 51(2), 205–231

Observatory of Economic Complexity (2025). https://oec.world/en/profile/bilateral-country/nga/partner/aus

Ramos TB and Caeiro S (2010). Meta-Performance Evaluation of Sustainability Indicators. *Ecological Indicators* 10(2), 157–166

Rinne J, Lyytimäki J and Kautto P (2013). From Sustainability to Well-Being: Lessons Learned from the Use of Sustainable Development Indicators at National and EU Level. *Ecological Indicators* 35, 35–42

Rydin Y (2007). Indicators as a Governmental Technology? The Lessons of Community-Based Sustainability Indicator Projects. *Environment and Planning D: Society and Space* 25, 610–624

Soroka SN (2006). Good News and Bad News: Asymmetric Responses to Economic Information. *The Journal of Politics* 68(2), 372–385

Thaler RH and Sunstein CR (2008). *Nudge: Improving Decisions About Health, Wealth, and Happiness*. Yale University Press, New Haven

Thaler RH and Sunstein CR (2021) *Nudge: The Final Edition*. 3rd edition. Penguin Random House, London

Turnhout E, Hisschemoller M and Eijsackers H (2007). Ecological indicators: Between the two fires of science and policy. *Ecological Indicators* 7, 215–228

Witting A (2017). Insights from 'Policy Learning' on How to Enhance the Use of Evidence by Policymakers. *Humanities and Social Sciences Communications*, Article number: 49

6

THE FUTURE FOR PERCEPTION-BASED INDICATORS OF SUSTAINABILITY

6.1 Introduction

Whether we like it or not, indicators have long been an important feature of sustainability and are likely to be with us for many years to come. The attraction of being able to capture complexity, a feature that is endemic to sustainability, with single numbers, a suite of numbers or even a short narrative, has much going for it when trying to convey information to non-experts. For managers, they provide the basis for planning as well as monitoring the attainment of targets, but they can also serve to help people conceptualise and frame what sustainability means for them and maybe even act as "nudges" to encourage action and behaviour change. For the media and the public, indicators can be used as ingredients within stories that set out aspects of the world we live in. People lead busy lives, and there seems to be an almost endless growth in availability of tools to help them gain access to information as quickly and as concisely as possible, and the revolution in artificial intelligence is perhaps the latest manifestation of that demand. It is often said that the need for "bite sized" chunks of information is almost insatiable, and indicators play very well into that dynamic.

The complex and multi-faceted nature of sustainability is the ideal context for indicators, even if the result is a list comprising hundreds of indicators as with the SDGs. The latter is much more of a catalogue or collection of indicators, which is the simplest approach that can be taken with these tools, and there are numerous alternatives, which try and provide indicators within a more systemic and cause–effect type of framework, which sets out the interactions between them. The DPSIR framework is one approach to the latter, but it also has its limitations as covered in Chapter 2. Some even try to combine all these indicators into a single number—an overall index of sustainability—but these can also have issues (Morse, 2019).

DOI: 10.4324/9781003456605-6

Indicators also have challenges as much can depend on how they are defined, framed, populated with values, communicated and interpreted, and changing any of these can result in varying conclusions. They are, after all, human constructs and are not the products of any laws of nature. It is this quintessential humanity of indicators that makes them fascinating, potentially useful but also potentially damaging. The technical details behind indicators, their methodology and mathematics, are merely reflections of that humanity. But what about perception-based indicators? What is their future, and does it provide some distinctive features when compared to the wider universe of SIs? That is the theme for this chapter.

6.2 A place for perception-based indicators of sustainability

The case is made in Chapter 2, using the set of SDG indicators for examples, that there are three broad types of SI based, in essence, on how the data to populate them have been collected which itself is just a reflection of what the indicator is trying to capture. We can think of this as in terms of a spectrum as shown in Figure 6.1, with the three types of indicator set out as distinct categories, although in practice, there will be some overlap between them; the boundaries between these categories will be fuzzy. At the two extremes are what can be called "reality-based" (or fact-based) and "perception-based" indicators. Those that are perception-based are unashamedly just that; the indicators are populated with values obtained from questionnaires using, for example, Likert scale-type responses, or perhaps interviews and participatory methods. The goal of the creators of these indicators is to capture perception because they think it matters! Maybe the thinking here is

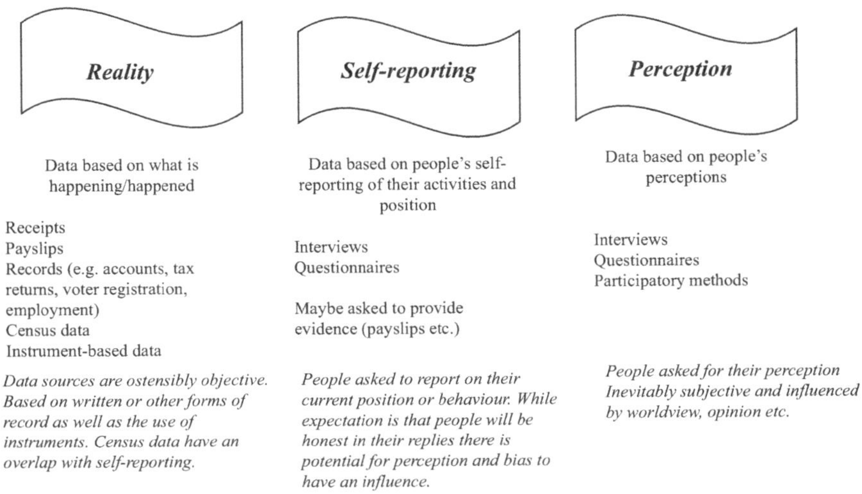

FIGURE 6.1 Three different types of indicators based on sources of data used to populate them

that perceptions can influence behaviour (e.g., perceptions of inflation or crime) or perhaps perceptions can provide a means for exploring "hard to reach" topics such as corruption. It is inevitable and indeed accepted at this end of the scale that people's perceptions will be influenced by a host of factors, and there is also likely to be much bias, both conscious and unconscious, at play. These are perhaps the most human of all indicators; they are designed to try and capture aspects of our lives and living. However, as set out in Chapter 2, the perception-based indicators comprise just 4% of the 247 SDG indicators and thus represent a small minority.

The other end of the spectrum comprises the bulk (85%) of the 247 SDG indicators, and here the sources are written, or other, records of activity—what has happened and what is happening? At this end of the spectrum, the data sources appear to be more objective and include readings from instruments and data collected from written records such as accounts, employment and registers. The records are seemingly "hard" things—things we can touch and dissect—and things that are "real" rather than being based on views or opinions—things we can have confidence in. At this end of the spectrum, it would seem that there is no potential for human failings such as bias to enter the scene, yet, as shown in Chapter 2; people are by no means absent. It is still people who make the decisions about how data are collected and how these data are manipulated to populate the indicators. These decisions can have a significant impact on the messages that emerge.

The third category, the one in the middle of Figure 6.1, is perhaps the most challenging to define. It has been labelled in Figure 6.1 as "self-reporting" and covers a swath of approaches, usually based on surveys, where people are asked to report on their status (e.g., employment, salary level, being a victim of crime) and their activities (e.g., level of smoking, alcohol consumption). This category is also relatively uncommon in the SDG framework, comprising just 11% of the SDG indicators. At one level, this category would seem to have much in common with indicators at the "reality" end of the spectrum. After all, people are being asked to self-report on their status and activities, and surely, these equate to reality? However, there is also a degree of potential overlap between self-reporting and perception-based indicators. Self-reporting may have a degree of bias, and people may under- or over-estimate aspects of their status or behaviour, or for some reason, they may under- or over-report income or other facets of their life. These biases, whether conscious or unconscious, may be influenced, for example, by an individual's perception of what is (or what is not) socially acceptable. If an individual perceives that society frowns upon smoking, alcohol consumption or taking drugs, or perhaps even driving petrol-fuelled cars, then it is possible that they may under-report these activities. Similarly, if the social norm is perceived to be the consumption of less meat, then people may bias their responses to reflect that. Hence, while people are not being asked directly to provide a perception, their responses can still be influenced by perception. There may also be overlap between the "real" and "self-reporting" categories with instruments such as the census which is based on self-reporting although there may be a legal requirement to complete it and provide correct information. The same is also true for self-reporting when it comes to tax and voter registration. This is why these specific instruments have been included under "Reality", but there is no guarantee that all these records are accurate.

Of the spectrum in Figure 6.1, perception-based indicators may sometimes be perceived as inferior compared to "Real" indicators; in effect, there can perhaps be an assumed gradient in terms of inherent quality from the left (high quality) to the right-hand side (low quality) of the spectrum. Having indicators that rely on people's perceptions of the world may inherently feel weak and subjective, given that perception-based indicators are open to all sorts of biases and delusions. Indeed, perceptions are often regarded as a type of proxy indicator where these are defined as (emphases provided by me):

> Also known as an indirect indicator. It is an *indirect sign or measure* that can *approximate* or can be *representative* of a phenomenon without the presence of a direct sign or measure.
>
> (The Law Dictionary)

> A proxy indicator acts as a *stand-in* to measure something we can't directly observe. Instead of measuring the real thing, we look at related signs that suggest its state or level. For example, using literacy rates as a stand-in for education quality. It's a shortcut that gives us an idea when direct data isn't available or is difficult to collect. Think of proxy indicators as *clues in a mystery*.
>
> *(Byte Benz, 2025)*

It is important to recognise that proxy indicators span the range of types shown in Figure 6.1. For example, the presence of capitals (e.g., car, house, TV) is of course readily measurable and can be regarded as a proxy measure of someone's income if they are reluctant to provide that information directly. Similarly, infant mortality rate can be measured—it is a "reality-based" indicator—and can be used as a proxy for the quality of the healthcare system. Hence, a proxy indicator does not have to be based on perception, but looking at it the other way around, perception-based indicators are often viewed as proxies. For example, the CPI may be seen by some as a proxy measure of corruption rather than a "real" measure and is thus open to criticism.

Nonetheless, I do not wish to create a "straw man" argument here, especially as there is much literature that recognises the value and need for perception-based indicators alongside the other types, as in Figure 6.1. Just as there are few who appear to argue for the replacement of reality-based indicators with those that are perception-based, then vice versa is also true. But this does bring us back to a question raised towards the end of Chapter 2—why are the perception-based indicators so poorly represented in the SDG framework? After all, as sustainability is inherently a human-centric concept by definition, it is after all about us and our relationship with the world around us and how that can influence the lives of future generations, then perceptions are clearly important, and this information needs to be valued in and of itself rather than just as a stop-gap while waiting for better, more reality-based, indicators to emerge. So, having just 4% of the SDG indicators being perception-based seems disappointing; are perceptions of so little value? It is certainly not a lack of these indicators that is the issue, and Table 3.1 is just a small sample of those that are available. I do not wish to promote any indicator, but if we take SDG 13 (take urgent action to combat climate change and its impacts), there are five targets and eight indicators, with the latter summarised as follows (emphasis from me):

- **Number** of deaths, missing persons and directly affected persons attributed to disasters per 100,000 population.
- **Number** of countries that adopt and implement national disaster risk reduction strategies in line with the Sendai Framework for Disaster Risk Reduction 2015–2030
- **Proportion** of local governments that adopt and implement local disaster risk reduction strategies in line with national disaster risk reduction strategies
- **Number** of countries with nationally determined contributions, long-term strategies, national adaptation plans, strategies as reported in adaptation communications and national communications
- **Total** GHG emissions per year
- Extent to which (i) global citizenship education and (ii) education for sustainable development are mainstreamed in (a) national education policies; (b) curricula; (c) teacher education; and (d) student assessment
- **Amounts** provided and mobilised in US$ per year in relation to the continued existing collective mobilisation goal of the $100 billion commitment through to 2025
- **Number** of least-developed countries and small island developing states with nationally determined contributions, long-term strategies, national adaptation plans, strategies as reported in adaptation communications and national communications

Most of these are clearly at the reality-indicator end of the spectrum in Figure 6.1, and not one of these is perception-based even though there are indicators such as the Climate Perceptions Index created by SPI and populated with data derived from Facebook, a social media platform with the aim to "provide insights for political leaders on where to focus to improve public support for climate action in their country" (Htitich et al., 2022; page 1). Hence, the Climate Perceptions Index was designed to capture people's perceptions to help inform action, and this is surely an important concern for all the SDG 13 indicators listed above both at national and local government scales. I need to stress, again, that I am not advocating as such for this indicator as it does have its limitations; the dependence on "Facebook" for arriving at the perceptions does raise issues, and this is, in fairness, readily acknowledged by SPI:

> Firstly, the sample only includes self-selected respondents who are the users of the Facebook platform. Secondly, we also need to be cautious about people's self-assessment where respondents often overestimate their own knowledge. And lastly, it is also important to note that with 107 countries covered by the survey, different questions will have different interpretations in different countries, stemming from diverse national and cultural contexts. This might lead to unexpected results on the question level. The survey likely under-represents those with lower levels of education and literacy, which would be reflected in the index scores.
>
> *(Htitich et al., 2022; page 13)*

But is it not important to compliment the indicators in SDG 13 with some that do capture people's perceptions of climate change, given how important they are when it comes to behaviour?

6.3 Perception-based indicators: The good, the bad and the beautiful

It goes without saying that indicators have benefits. If they did not, then no one would want them. Their ability to capture complexity into a single value, often paired with powerful visual devices such as graphs, "traffic lights", "Dashboards" and "league-Table" style ranking, allow for those who are not experts in the issue covered by the indicator to get a sense as to what is happening; are matters good or bad and are they worsening or improving? They may tell us very little, if anything, about why things are the way that they are, but they do provide a means to capture the world around us and communicate that information to those who can make decisions. Perceptual indicators also have this benefit of capturing complexity and aiding communication. Are perception-based indicators useful in the sense that they are more "human" and address issues that are "true to our hearts" such as happiness? Well, the answer is clearly yes, as otherwise they would not exist, and the evidence from Chapter 3 is that there are many of them. It is true that perception-based indicators do not feature that much within the SDG framework, but they are certainly abundant within what could be called the "sustainability indicator multiverse"—the collection of all indicators, how they relate to each other and how they are communicated and used.

In terms of the benefits of perception-based indicators, there are certainly some issues that can best be captured by using perception. After all, if we are interested in what people think about a topic, such as their views on climate change, then there is no alternative but to ask them. Perception-based indicators are also useful in circumstances when it can be challenging to get direct (real) measures, and a case in point here is corruption (Sampford et al., 2006; Heywood and Rose, 2014). As with assessing crime in general, there are various approaches that could be taken. Following the "reality" end of the spectrum in Figure 6.1, we could collect data by observing the incidence of corruption or by relying on reported incidences to the authorities or, better still, successful prosecutions. But there are issues here as these statistics may provide only a partial picture; the tip of the iceberg so to speak. Observing corruption is difficult, to say the least, given its hidden nature, and there may well be safety issues at play when it comes to collecting such data. People can be afraid to report incidents of corruption as they may think there will be retribution. Indeed, it could be that the police and others in the judicial system are also corrupt. After all, people may be paying for services (education, healthcare, etc.) that they should receive free of charge from a provider, be it government or a company, and retribution can take the form of barring access to these services or perhaps even a demand that they pay more. All of this can mean that recorded incidents and successful prosecutions of corruption may be low. In the "self-reporting" category of Figure 6.1, we can try asking people, with assurances of anonymity, to catalogue their own personal experiences of corruption. However, this requires much trust given the potential risks for the respondent and is open to the biasing issue noted above, which means that people may over- or understate their experiences. They may have various motivations behind the latter, including, of course, issues with memory recall. Given all these issues, the perception approach to assessing corruption has its advantages. Here, people are asked for their views regarding the extent of corruption in society rather than being asked to provide their own personal experiences, and respondents may feel safer as they are talking in general terms with nothing that can be traced back to them.

A dilemma with the perception approach to corruption is that this can be influenced by a host of factors, some of which are shown in Figure 6.2. The figure has a horizontal axis with so-called "reality" indicators to the left-hand side and perception-based indicators to the right-hand side. Indicators based on self-reporting of experience and behaviour are towards the centre of the axis. At the reality end of the spectrum are records of incidents, prosecutions and convictions linked to corruption. These are often reported in the media and will help frame perceptions of the incidence of corruption within their readership. Similarly, perceptions of corruption can also be framed by personal experience, biases, worldviews of social and cultural norms and the views of family and friends, who could in turn be influenced by reporting in the media and elsewhere (e.g., social media). Reporting of incidents of corruption by the media, including social media, is likely to be important but may also be influenced by factors such as the political and business leanings of media owners, editors, journalists and maybe even advertisers. Hence, while there has been research designed to explore how perceptions of corruption can be influenced by the media, including social media (Charron and Annoni, 2021), the influence of media owners, some of whom may themselves

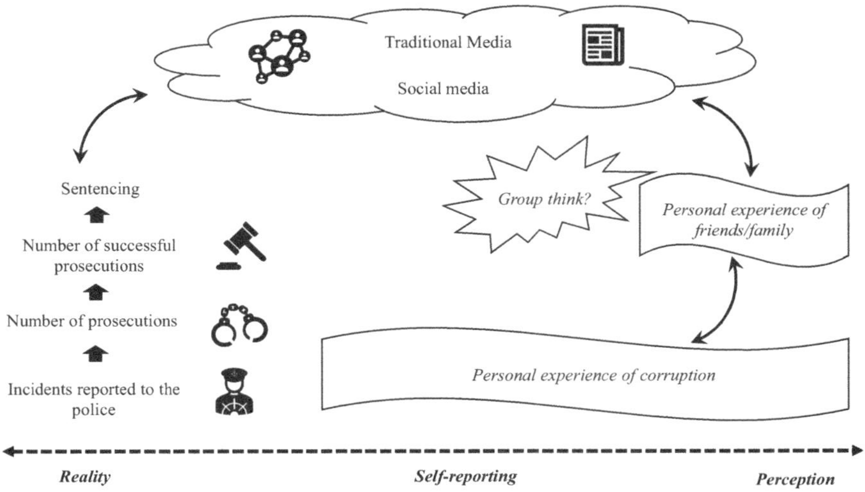

FIGURE 6.2 Some potential ways of assessing the prevalence of corruption in society using indicators

be politicians, on the media reporting of corruption remains relatively underexplored. The wider influence of newspaper owners on reporting in the publications they own has been well-reported (Kedia and Kim, 2024), and a recent review and analysis is provided by Benson et al. (2025), but how they influence reporting of corruption needs much more elucidation.

High-profile court cases widely reported by the media can potentially have a large impact in framing perceptions even though these may be unrepresentative of what is happening in wider society. There can be echo-chamber effects here and a sort of "group think" as these high-profile cases, amplified by the traditional and social media and others, can lead to perceptions that may simply be unwarranted (Gilman, 2018). A similar effect can occur, as individual experiences may become magnified via social media, leading to a "group think" that corruption must be highly prevalent when perhaps it is not. Indeed, what some societies see as corruption may be regarded by others as an acceptable practice, and much can depend on the definition (Maciel et al., 2024). For example, the CPI is based on data derived via a set of surveys designed to explore perceptions of corruption of countries visited by respondents, with the latter being primarily businesspeople but also journalists and academics (de Maria, 2008; Donchev and Ujhelyi, 2014; Zouaoui et al., 2017; Gilman, 2018; Baumann, 2020). Some in the sample may be resident in the county they are assessing, while others will base their perceptions on visits—perhaps relatively short ones. Indeed, given an interest from the business community in the assessment of corruption, it is perhaps unsurprising that some of the perception surveys used to calculate the CPI are undertaken by commercial organisations that sell their findings (Baumann, 2020). Unfortunately, perceptions can become resilient and hard to change and may result in a sense that corruption is part of the "culture" of a society and generate a misleading view that some societies are inherently more prone to it than are others (World Bank, 1997; Mujtaba et al., 2013). Therefore, an outcome arising from a country doing badly in the CPI rankings could well be a loss of investment and trade; attracting a bad perception can be costly!

The reputational power of the CPI and its impact on business investment and growth is probably a major reason for its reporting by the media as shown in Chapter 5. Happiness may be of interest to all of us and can even be a central concern when it comes to developing policy as shown in Bhutan, but a perception that corruption is rampant in a country can potentially damage the economy. Not only will companies wish to avoid the additional costs of doing business in a corrupt country, but there are also serious risks of litigation to worry about, as illustrated with the newspaper reporting from Australia covered in Chapter 5. It is no wonder that those in government in countries which do badly in the rankings are quick to dismiss the CPI even if their arguments can seem tortuous as shown in Chapter 5 with media reporting of the CPI linked to Nigeria. The argument that people may perceive corruption as being high in Nigeria (hence the poor CPI) because of a low conviction rate, and presumably vice versa, seems somewhat contrived to say the least, and the available evidence from research that has explored the connection between convictions for corruption and people's perceptions suggest the relationship is far more complex (Pan et al., 2023).

What about the disadvantages of perception-based indicators? Some of these have already been discussed throughout the book, including the potential for bias and influence from a range of factors, as illustrated above for the CPI, and as with all indicators based on collecting data from people, there is the ever-present issue of sampling. Methodological challenges surrounding the use of tools such as the Likert scale have been covered in Chapter 3, but arguably, the most important disadvantage of perception-based indicators is the same for indicators in general and is, ironically, the mirror image of their advantage. If indicators are intended to capture complexity and reduce it to single numbers, there is always the danger that much valuable information will be missed. The result can be a tendency to "manage to the indicators" to the exclusion of anything else. In effect, indicators can choke the emergence of a richer narrative by restricting the story to that framed by the indicators and nothing more (Bell and Morse, 2003, 2008). Indicators can become institutionalised, and people may resist any attempt to think more broadly "outside the box". A multi-indicator framework, such as that represented by the SDGs, is intended to help counter this criticism, and in sustainability, this point is especially resonant given how, by definition, it spans environmental, social and economic dimensions, and these are very broad indeed. However, there are obviously limits as to how many indicators can be included in any framework before it becomes unwieldy. Would it be a good idea, for example, to expand the SDG indicator list to thousands? This may generate a much richer story, but how workable would such a framework be in practice?

As noted above, and taking care not to fall into a "straw man" argument, attempts to include indicators based on perceptions alongside those based on "reality" (Figure 6.1) can potentially result in a certain bias when it comes to valuing the indicators. I have no empirical evidence to draw on here—just personal experience. In my view, some of the criticism I have heard of perception-based indicators flows from a mistaken tendency to see them as being promoted as replacements for indicators that sit towards the "reality" end of the spectrum. I have never heard anyone propose that, but I guess it may be linked to a suspicion that the hard work and cost of development and populating a reality-based indicator is being avoided by opting for something based on perception that is easier and cheaper to do. But surely, the value of perception-based indicators is their ability to be complementary to other types of indicators rather than replace them; perception-based indicators can only add to the story rather than diminish it.

6.4 Thoughts on the future of perception-based indicators of sustainability

All indicators, including those developed for sustainability, have a degree of subjectivity associated with them, yet at the same time, there is often much trust in those that develop these tools, and a foundational assumption that they provide objective and unbiased pictures of the world. As noted in Chapter 5, one of the reasons that indicators produced by highly respected bodies such as the UN and indeed non-government organisations such as TI are so readily reported by the media is the trust associated with them. In fairness to those who produce and curate indicators, the methods they employ are nearly always given along with the data sets that they use, but the casual consumer is unlikely to take the time needed to dissect these and appreciate the ramifications that emerge out of assumptions made to get from raw data to published indicators and any rankings based on them. But all indicators, even those at the reality end of the spectrum in Figure 6.1, are human constructs designed to help us create a picture of the world. Hence, the GDP, the archetypal objective indicator of economic activity which is a component of many frameworks and indeed other composite indicators such as the HDI, captures just some elements of the economy and omits others. Alternatives to the GDP such as the GPI (Berik, 2020; Jansen et al., 2024) and the ISEW (Tsara et al., 2024) were created to address what some saw as failings with GDP, but they have their own issues including the availability of good quality data over time. While indicators such as the GDP, GNP and GNI have certainly come in for much criticism, perhaps understandable, given their strong influence on policy makers and others, they are not in and of themselves bad indicators. GDP was designed to capture the size of an economy based on flows of money, and that is what it and its relatives do (Coyle, 2014), nothing more and nothing less. There is certainly plenty of scope for critiquing the ways in which economic data are collected (Coyle, 2025), but the GDP was not designed to include environmental and social costs or encompass a judgement call on the benefits of goods and services, and neither was it intended as an indicator of well-being or quality of life.

Perhaps, there are various possible futures for perception-based indicators. First, if they are seen as merely being stepping stones to the achievement of what some perceive to be "better" (i.e., more real or objective) indicators, then we can imagine a future where perception-based indicators are gradually replaced as instrument-based and other "better" systems of recording data improve. In this scenario, which can perhaps be regarded as "practical tolerance", perception-based indicators are not regarded as having any intrinsic value and are merely stop-gaps on the road to getting indicators that are at the "reality" end of the spectrum in Figure 6.1. A second possible future, which can perhaps be categorised as "unmissable value", is where perception-based indicators are accepted as a necessary and valuable device for engaging people within sustainability. In this scenario, there will always be a need for perception-based indicators not because of failings in our ability to construct more reality-based indicators, but because sustainability is about people and we engage with the world in many ways, and much of that engagement is framed by our perceptions of it. Hence, given that the attainment of sustainability inevitably involves a need to engage with people, maybe by shifting expectations and behaviour, there will always be a need to capture and understand perspectives.

But which of these scenarios is the more likely? Well, based on the dominance of "reality" indicators in the SDG framework (85% of indicators) with just 4% of them classified as being perception-based, it would seem that the centre of gravity in the indicator multiverse is arguably much more towards the "reality" end of the spectrum in Figure 6.1. Perception-based indicators do seem to have a niche within our current multiverse of SIs, but that niche would seem so far to be a relatively small one. Hence, the (albeit) limited evidence to date perhaps leans towards the "practical tolerance" hypothesis rather than "unmissable value", with the SDG framework having only a sprinkling of perception-based indicators. While the list of examples set out in Table 3.1 is a long one, how many of these have really found their way into widely accepted SI frameworks? It appears that even happiness is excluded from the SDGs! Maybe we need to be sustainable but not necessarily happy.

One of the intriguing questions associated with perception-based indicators is whether they have an innate appeal to people that is greater than the reality type of indicators in Figure 6.1. Given the very human basis of the perception-based indicators and the kind of topic areas (crime, happiness, sense of place, etc.) that they capture, is there any evidence that they attract more interest than the reality type of indicator that may often seem more technical and less human? It is certainly true that perception-based indicators readily emerge out of a participatory process, and an example is provided in Chapter 4 for a series of SSM workshops held in Malta. Some of the participants in this workshop did feel a need to include perception-based indicators, such as indicators designed to capture a sense of media reporting of tourism in Malta, but the same participants included indicators that could perhaps be regarded more at the reality end of the spectrum in Figure 6.1. There was no evidence from these workshops for the focus to be solely or even majorly on perception-based indicators, but participants clearly saw some value in having them.

In Chapter 5, the extent of media reporting for perception-based indicators was compared to those that are not perception-based, such as the GDP, HDI and EF, and frankly, there was no evidence here of any media emphasis on the perception-based type of indicator. The only perception-based indicator that attracted much in the way of newspaper interest was the CPI, but the extent of reporting was still less than that for the HDI. Indicators of happiness attracted far less media attention than did the HDI or the EF. This is not to say that perception-based indicators are not used by the media, far from it, but there is no evidence that they are favoured compared with some of their non-perception-based cousins. But this is a fair comparison as perhaps what matters most is the topic area that the indicator is designed to address and the nature and reach of the organisation promoting the indicator rather than the indicator per se. Indeed, of all the indicators looked at in Chapter 5, the clear winner in terms of the extent of its use in the media was the GDP. But maybe this is not because GDP is somehow a "better" indicator than all the others, but that it exists in a space, economic performance of nation states, that is of such wide interest to politicians, businesspeople and others with power. Also, the use of indicators by the media and resonance with people can be very different things, and this is a topic that would certainly warrant much more research. Great care does need to be taken when comparing the media reporting of indicators and drawing conclusions.

Going beyond an exploration of indicator use by the media, there is very little research designed to explore the influence of such indicators on policy and decision-making. This may perhaps be surprising given that many see SIs in an instrumental sense, as devices to help guide interventions. But one of the issues here is that it can be challenging to unpick the influence of an indicator relative to all the other influences that may exist, including those from other indicators. Can a particular decision really be tied to an indicator or suite of indicators, or is the influence much more subtle and complex with many factors such as awareness, learning and nudging at play? There are various frameworks designed to help classify various uses of indicators, but this is a very complex and fuzzy area to explore, and there are dangers of assuming direct cause and effect. Given the relatively underdeveloped nature of our understanding of the use and influence of indicators, are there any a priori reasons to imagine that perception-based indicators may fare better than their more reality-based cousins? There are solid examples of influence, and the CPI is one of them as illustrated by the analysis of its media reporting in Nigeria and Australia set out in Chapter 5. It appears that the

CPI does attract the attention of those in and not in government, even if that attention largely takes the form of attempts to discredit the indicator. Even attempts to discredit the CPI serve to increase attraction and awareness of the very issue that the indicator was designed to address. But are these stories of influence more prevalent or stronger for the CPI compared with other indicators not based on perception? The limited evidence available to date would not seem to support such a claim (Morse, 2023).

6.5 The verdict is in

So, how to end this book? Indicators of all the shades shown in Figure 6.1, be they reality, self-reporting or perception-based, generate all sorts of tensions. On the one hand, they are the products of people for other people to use, and the intention is for them to help us manage complexity. SIs are no different in this regard. Perception-based indicators may be a minority within the SDG framework and perhaps even within the wider indicator multiverse, but they have a place that is not just about them being stepping-stones until something better comes along. Seeing perception-based indicators as nothing more than "practical tolerance" may be a common view, but I think it is far too narrow and misses the value and opportunity that perception-based indicators provide—the representation of the views of people, be it their wishes, wants, desires and their perceptions of the world around them and why it is the way it is. They are not merely stand-ins to the main act but valuable in themselves. I personally much prefer to see them as representing "unmissable value", and it is disappointing to see them so under-represented in the SDG list of indicators.

This book has made the case that perception-based indicators of sustainability have an intrinsic and "unmissable" value in themselves, and we are perhaps missing an opportunity, not least within the SDG framework. Whatever emerges as the follow-up to the SDGs should have a solid presence of perception-based indicators across all the goals. It would help inculcate a sense of lateral thinking and creativity within sustainability—looking at issues from a range of different perspectives, including those who are meant to be part of the solution. Edward de Bono, the originator of lateral thinking, puts it very well in his book on creativity:

> Most of the mistakes in thinking are inadequacies of perception rather than mistakes of logic. The astonishing thing is that traditionally we have always put the emphasis on logic rather than on perception. We have felt uncomfortable with the fluidities and "possibilities" of perception and have taken refuge in the apparent certainties and "truth" of logic. This has served us well when the input is as definite as a measurement or a number but much less well in those areas where the excellence of processing cannot make up for the inadequacies of perception.
>
> *(de Bono, 1992; page 58)*

Maybe there are lessons here for all of us who work within the multiverse of SIs. Have we sought to take refuge in what we think are indicators which provide us with truthful (i.e., real) windows on our complex and messy world? Perhaps, we need to have the creativity, courage and conviction to bring happiness, amongst much else we value so deeply, to the heart of sustainability.

6.6 Summary

Perception-based indicators comprise very few of the SDG indicators, and indeed, they are not as prevalent as some of the non-perception-based indicators in newspaper reporting. Given that perception-based indicators are typically designed to capture people's stances on very human traits, such as happiness and well-being, this may be surprising. Maybe part of the reason for this is an implied, maybe even unconscious, hierarchy within the indicator community, which favours more "reality"-based indicators—those that can be populated based on instrument readings and records. Perception-based indicators may be seen by some as just stepping-stones—something to fill a gap and put up with until better, more reality-based indicators can emerge. But they are more than that and should be seen as having value in themselves and not just as stepping stones to a better destination.

6.7 Food for thought

1. Please come up with perception-based indicators to supplement those in the UN's SDG indicator list and explain your choices.
2. Why do perception-based indicators not attract more attention in the media?
3. What is the case for not having perception-based indicators in sustainability frameworks such as the SDGs?
4. Can the DPSIR framework of indicators be applied to perception-based indicators? Please provide examples to illustrate your answer.
5. Please come up with other versions of Figure 6.2 for topics of your choice other than corruption.
6. How would you explore the resonance that perception-based indicators may have with people compared to their more "reality-based" cousins?

References

Baumann H (2020). The Corruption Perception Index and the Political Economy of Governing at a Distance. *International Relations* 34(4), 504–523

Bell S and Morse S (2003). *Measuring Sustainability. Learning by Doing*. Earthscan, London

Bell S and Morse S (2008). *Sustainability Indicators. Measuring the Immeasurable*. 2nd edition. Earthscan, London

Benson R, Hessérus M and Neff T (2025). *How Media Ownership Matters*. Oxford University Press, Oxford

Berik G (2020). Measuring What Matters and Guiding Policy: An Evaluation of the Genuine Progress Indicator. *International Labour Review* 159(1), 71–94

Byte Benz (2025). *What Is a Proxy Indicator? A Comprehensive Guide to Understanding Its Role in Data Analysis*. Available at: https://medium.com/@bytebenz/what-is-a-proxy-indicator-a-comprehensive-guide-to-understanding-its-role-in-data-analysis-7323a823532d

Charron N and Annoni P (2021). What Is the Influence of News Media on People's Perception of Corruption? Parametric and Non-Parametric Approaches. *Social Indicators Research* 153, 1139–1165

Coyle D (2014). *GDP. A Brief but Affectionate History*. Princeton University Press, Princeton, NJ

Coyle D (2025). *The Measure of Progress: Counting What Really Matters*. Princeton University Press, Princeton, NJ

De Bono E (1992). *Serious Creativity: How to be Creative Under Pressure and Turn Ideas into Action*. Harper Business, New York

De Maria B (2008). Neo-Colonialism Through Measurement: A Critique of the Corruption Perception Index. *Critical Perspectives on International Business* 4(2/3), 184–202

Donchev D and Ujhelyi G (2014). What Do Corruption Indices Measure? *Economics & Politics* 26(2), 309–331

Gilman SC (2018). To Understand and to Misunderstand How Corruption Is Measured: Academic Research and the Corruption Perception Index. *Public Integrity* 20, S74–S88

Heywood PM and Rose J (2014). "Close But No Cigar": The Measurement of Corruption. *Journal of Public Policy* 34(3), 507–529

Htitich M, Krylova P, Harmacek J and Lisney J (2022). *Climate Perceptions Index. Social Progress Imperative*. Available at: www.socialprogress.org

Jansen A, Wang R, Behrens P and Hoekstra R (2024). Beyond GDP: A Review and Conceptual Framework for Measuring Sustainable and Inclusive Wellbeing. *Lancet Planet Health* 8, e695–e705

Kedia S and Kim G (2024). Impact of Media Ownership on News Coverage. *Management Science* 70(9), 5627–5645

Maciel GG, Magalhães PC, de Sousa L, Pinto IR and Clemente F (2024) A Scoping Review on Perception-Based Definitions and Measurements of Corruption. *Public Integrity* 26(1), 114–131

Morse S (2019). *The Rise and Rise of Indicators: Their History and Geography*. Routledge, London

Morse S (2023). *Ranking Nations: The Value of Indicators and Indices*. Edward Elgar, Cheltenham, Gloucestershire

Mujtaba BG, Williamson P, Cavico FJ and McClelland B (2013). Managing the Link Between Bribery and Wealth Based on Corruption Perception Index (CPI) and Gross National Product (GNP) per Capita. *Journal of Management Policy and Practice* 14(3), 100–115

Pan Y, Shu Z and Ye Z (2023). Exploring the Dynamics of Corruption Perceptions in Sustained Anti-Corruption Campaigns: A Survey Experiment in China. *Humanities and Social Sciences Communications* 10, Article number: 570

Sampford C, Shacklock A, Connors C and Galtung F (eds) (2006). *Measuring Corruption*. Ashgate, Aldershot, UK

Tsara IK, Vortelinos DI and Menegaki AN (2024). The Index of Sustainable Economic Welfare (ISEW) as a Proxy for Sustainable GDP: Revisited and Recapitulated. *Discover Sustainability* 5, 158

World Bank (1997). *World Development Report 1997. The State in a Changing World*. Oxford University Press, New York

Zouaoui A, Al Qudah A and Arab MB (2017). World Corruption Perception Index Analysis. *Research Journal of Finance and Accounting* 8(24), 85–91

INDEX

Note: Page numbers in *italics* indicate figures, and page numbers in **bold** indicate tables in the text.